Studies in Systems, Decision and Control

Volume 188

The series "Studies in Systems, Decision and Control" (SSDC) covers both new developments and advances, as well as the state of the art, in the various areas of broadly perceived systems, decision making and control–quickly, up to date and with a high quality. The intent is to cover the theory, applications, and perspectives on the state of the art and future developments relevant to systems, decision making, control, complex processes and related areas, as embedded in the fields of engineering, computer science, physics, economics, social and life sciences, as well as the paradigms and methodologies behind them. The series contains monographs, textbooks, lecture notes and edited volumes in systems, decision making and control spanning the areas of Cyber-Physical Systems, Autonomous Systems, Sensor Networks, Control Systems, Energy Systems, Automotive Systems, Biological Systems, Vehicular Networking and Connected Vehicles, Aerospace Systems, Automation, Manufacturing, Smart Grids, Nonlinear Systems, Power Systems, Robotics, Social Systems, Economic Systems and other. Of particular value to both the contributors and the readership are the short publication timeframe and the world-wide distribution and exposure which enable both a wide and rapid dissemination of research output.

More information about this series at http://www.springer.com/series/13304

Antonio André Novotny
Jan Sokołowski · Antoni Żochowski

Applications of the Topological Derivative Method

Springer

Antonio André Novotny
Coordenação de Métodos
Matemáticos e Computacionais
Laboratório Nacional de Computação
Científica LNCC/MCTIC
Petrópolis, Brazil

Jan Sokołowski
Institut Élie Cartan de Nancy, UMR 7502
Université de Lorraine, CNRS
Vandœuvre-Lès-Nancy, France

and

Systems Research Institute
Polish Academy of Sciences
Warsaw, Poland

and

Universidade Federal da Paraiba, Centro de informatica
João Pessoa, PB, Brazil

Antoni Żochowski
Systems Research Institute
Polish Academy of Sciences
Warsaw, Poland

ISSN 2198-4182 ISSN 2198-4190 (electronic)
Studies in Systems, Decision and Control
ISBN 978-3-030-05431-1 ISBN 978-3-030-05432-8 (eBook)
https://doi.org/10.1007/978-3-030-05432-8

Library of Congress Control Number: 2018963050

This Springer imprint is published by the registered company Springer Nature Switzerland AG
The registered company address is: Gewerbestrasse 11, 6330 Cham, Switzerland

This book is dedicated to
Bożena, Renata, and Vanessinha

Foreword

The book is written by three well-known specialists in the shape and topology optimization. The book contains new results on applications of the topological derivative method in control theory, topology optimization, and inverse problems. The method is now of common use in the optimum design in mechanics. The topological variations of geometrical domains are useful for numerical solutions of inverse problems and in imaging. The variations of the optimal value of the cost for a class of optimal control problems are introduced. Thus, the new concept is useful also in systems sciences. The book can be recommended to Ph.D. students and researchers in applied mathematics, computational mechanics, and control theory.

Montreal, Canada
Michel Delfour

Preface

The topological derivative method is recognized as a robust numerical technique in engineering applications such as topology optimization, inverse problems, imaging processing, multi-scale material design, and mechanical modeling including damage and fracture evolution phenomena. In particular, the topological derivative is defined as the first-order term of the asymptotic expansion of a given shape functional with respect to a small parameter that measures the size of singular domain perturbations, such as holes, inclusions, defects, source terms, and cracks. Therefore, the method is based on the asymptotic approximations of solutions to elliptic boundary value problems combined with mathematical programming tools.

The book presents new results and applications of the topological derivative method in control theory, topology optimization, and inverse problems. The theory in singularly perturbed geometrical domains is introduced in Chaps. 1–5 through some selected examples. A first-order topology design algorithm based on the topological derivative concept is presented in Chap. 6 and further developed in Chaps. 7–9, together with its applications in the context of topology optimization. A second-order reconstruction algorithm of Newton type based on higher-order topological derivatives is proposed in Chap. 10 and presented in details in Chap. 11, which is specifically designed for solving a class of inverse reconstruction problems. The resulting algorithm is non-iterative and hence very robust with respect to noisy data and also independent of any initial guess.

The present work is complementary to the monograph by A. A. Novotny and J. Sokolowski, Topological Derivatives in Shape Optimization, Interaction of Mechanics and Mathematics Series, Springer-Verlag, Berlin, Heidelberg, 2013. In particular, it presents not only the theory, but also a wide portfolio of applications in engineering. Therefore, this book is oriented to researchers and students in applied mathematics and computational mechanics interested on the mathematical aspects of the topological asymptotic analysis as well as on applications of the topological derivative method in computational mechanics.

This book is the result of scientific collaboration between Argentina, Brazil, France, Poland, and Russia, as well as on the research work of former Ph.D. students in Nancy (France) and Petrópolis (Brazil). The collaboration between

Antonio André Novotny and Jan Sokołowski was supported by IECN in France and by CNPq, FAPERJ, and LNCC in Brazil. The research activities of Antoni Żochowski was supported by IBS PAN in Warsaw, Poland.

Petrópolis, Brazil
Nancy, France
Warsaw, Poland
September 2018

Antonio André Novotny
Jan Sokołowski
Antoni Żochowski

Contents

1 Introduction 1
1.1 Theoretical Framework for Local Solutions 2
1.2 Elementary Example of Topological Derivative 5
1.3 Shape and Topology Optimization 6
1.4 Evaluation of Topological Derivatives 8
1.5 Open Problems for Topological Derivative Method 10
1.6 Description of the Content of the Book 12

2 Theory in Singularly Perturbed Geometrical Domains 13
2.1 Preliminaries 13
2.2 Asymptotic Expansions for the Domain Decomposition Technique 14
2.2.1 Asymptotic Expansions of Steklov–Poincaré Operators 14
2.2.2 From Singular Domain Perturbations to Regular Perturbations of Bilinear Forms in Truncated Domains 16
2.2.3 Signorini Problem in Two Spatial Dimensions 19
2.2.4 Domain Decomposition Method for Elasticity 21
2.3 Matched Asymptotic Expansions for Neumann Problem 26
2.3.1 Asymptotic Expansion of the Steklov–Poincaré 26
2.3.2 Asymptotic Expansion of the Linear Form 29
2.3.3 Asymptotic Expansion of the Energy Functional 30
2.4 Asymptotics of Steklov–Poincaré Operators in Multilayer Subdomains 34
2.4.1 Multilayer Inclusions 34
2.4.2 Steklov–Poincaré Operator in Multilayer Inclusion 36
2.4.3 Asymptotic Expansions in Multilayer Subdomain 37
2.4.4 Multilayer Subdomains in Linear Elasticity 38

3 Steklov–Poincaré Operator for Helmholtz Equation ... 41
3.1 Representation of Solutions for Helmholtz Equation ... 41
3.1.1 Case I: Coercive Operator ... 42
3.1.2 Case II: Non-coercive Operator ... 44
3.2 Numerical Testing of Approximate Formulas for Steklov–Poincaré Operators ... 45
3.3 Solutions in the Ring for Helmholtz ... 47
3.4 Precision of Formulas for Helmholtz in Both Cases ... 49

4 Topological Derivatives for Optimal Control Problems ... 51
4.1 Example in One Spatial Dimension ... 51
4.2 Control Problem ... 52
4.3 Topological Derivative ... 54
4.4 Numerical Example ... 58
4.5 Final Remarks ... 58

5 Optimality Conditions with Topological Derivatives ... 61
5.1 Preliminaries ... 62
5.2 Model Problem ... 63
5.3 Double Asymptotic Expansion ... 67
5.4 Topological Differential with Respect to Multiple Holes ... 70
5.5 Dependence of Solutions on Boundary Variations ... 73
5.6 Simultaneous Topology and Shape Modification ... 78
5.7 Analytical Example ... 81

6 A Gradient-Type Method and Applications ... 85
6.1 Preliminaries ... 86
6.2 First Order Topology Design Algorithm ... 87
6.3 Shape and Topology Optimization ... 89
6.3.1 Structural Topology Design ... 90
6.3.2 Fluid Flow Topology Design ... 96
6.3.3 Multiscale Material Design ... 99
6.3.4 Additional Applications ... 103
6.4 Future Developments ... 105

7 Synthesis of Compliant Thermomechanical Actuators ... 109
7.1 Preliminaries ... 109
7.1.1 Simple Example of a Bar Structure ... 110
7.1.2 Topological Derivative for Inclusions ... 111
7.2 Problem Formulation ... 112
7.2.1 Unperturbed Problem ... 112
7.2.2 Perturbed Problem ... 114
7.3 Existence of the Topological Derivative ... 116
7.4 Topological Asymptotic Analysis ... 119

7.4.1 Contrast on the Elastic Coefficients 119
7.4.2 Contrast on the Thermal Coefficients 121
7.4.3 Topological Derivative 122
7.5 Numerical Experiments 123
7.5.1 Example 1: Amplifier 124
7.5.2 Example 2: Inverter with Eccentricity Effect 126
7.6 Final Remarks 128

8 Synthesis of Compliant Piezomechanical Actuators 129
8.1 Preliminaries 129
8.2 Problem Formulation 133
8.2.1 The Mechanical Model 134
8.2.2 The Shape Functional 135
8.2.3 The Adjoint State 135
8.3 Topological Derivative 136
8.3.1 Abstract Setting of the Problem 137
8.3.2 Topological Asymptotic Expansion of the Steklov–Poincaré Operator 139
8.3.3 Topological Asymptotic Expansion of the Solution . . . 140
8.3.4 Topological Asymptotic Expansion of the Shape Functional 141
8.4 Numerical Experiments 143
8.4.1 Example 1: Moonie 144
8.4.2 Example 2: Inverter 146
8.5 Final Remarks 147

9 Asymptotic Analysis of Variational Inequalities 149
9.1 Motivation 150
9.2 Problem Formulation 152
9.3 Domain Decomposition Technique 154
9.4 Topological Asymptotic Analysis 155
9.5 Topological Derivative Formula 160
9.6 Numerical Application 161
9.7 Final Remarks 163

10 A Newton-Type Method and Applications 165
10.1 Preliminaries 165
10.2 Higher Order Topological Derivative 166
10.3 Second Order Reconstruction Algorithm 167
10.4 Inverse Reconstruction Problems 170
10.4.1 Inverse Gravimetry Problem 171
10.4.2 Pointwise Source Reconstruction Problem 174
10.4.3 Obstacle Reconstruction Problem 177

10.4.4 Inverse Conductivity Problem 179
10.4.5 Electromagnetic Casting Problem 180
10.5 Future Developments 180

11 The Electrical Impedance Tomography Problem 183
11.1 The Inverse Conductivity Problem 183
11.2 Topological Asymptotic Expansion 186
11.2.1 Asymptotic Expansion of the Solution 187
11.2.2 Asymptotic Expansion of the Shape Functional 192
11.2.3 Introduction of Adjoint States 194
11.3 A Noniterative Reconstruction Algorithm 197
11.4 Numerical Experiments 198
11.5 Final Remarks 199

References 201

Chapter 1
Introduction

Mathematical analysis and numerical solutions of problems with unknown shapes or geometrical domains is a challenging and rich research field in the modern theory of the calculus of variations, partial differential equations, differential geometry as well as in numerical analysis. In this book, we describe some aspects of numerical solution for problems with unknown shapes, which use tools of asymptotic analysis with respect to small defects or imperfections to obtain sensitivity of shape functionals. In classical numerical shape optimization, the boundary variation technique is used with a view to applying the gradient or Newton type algorithms. Shape sensitivity analysis is performed by using the velocity method. In general, the continuous shape gradient and the symmetric part of the shape Hessian are discretized. Such an approach leads to local solutions, which satisfy the necessary optimality conditions in a class of domains defined in fact by the initial guess. A more general framework of shape sensitivity analysis is required when solving topology optimization problems. A possible approach is asymptotic analysis in singularly perturbed geometrical domains. In such a framework, approximations of solutions to boundary value problems (BVPs) in domains with small defects or imperfections are constructed, for instance by the method of matched asymptotic expansions. The approximate solutions are employed to evaluate shape functionals, and as a result topological derivatives of functionals are obtained. In particular, the topological derivative is defined as the first term (correction) of the asymptotic expansion of a given shape functional with respect to a small parameter that measures the size of singular domain perturbations, such as holes, cavities, inclusions, defects, source-terms and cracks. This new concept of variation has applications in many related fields, such as shape and topology optimization, inverse problems, image processing, multiscale material design and mechanical modeling involving damage and fracture evolution phenomena. In this chapter, the topological derivative concept is presented in detail within the framework of the domain decomposition technique. Such an approach is constructive, e.g., for coupled models in multiphysics as well as for contact problems in elasticity. Further in the book,

A. A. Novotny et al., *Applications of the Topological Derivative Method*,
Studies in Systems, Decision and Control 188,
https://doi.org/10.1007/978-3-030-05432-8_1

we describe the first and second order numerical methods of shape and topology optimization for elliptic BVPs, together with a portfolio of applications and numerical examples in all the above mentioned areas.

1.1 Theoretical Framework for Local Solutions

The parametric optimization is a challenging field of applied mathematics with many applications for the real life problems solution. The presence of the state equation makes the parametric optimization involved from mathematical point of view. We restrict our investigations to the elliptic boundary value problems. Usually such problems are ill-posed in the natural setting. Roughly speaking, the weak convergence of minimizing sequence of parameters does not imply the convergence of solutions to the state equation. The so-called G-convergence was introduced for the elliptic operators by Sergio Spagnolo in order to study the weak convergence of solutions, so it is an abstract convergence of the inverse operators. It is known that such an abstract convergence is not equivalent to any convergence of coefficients of operators in function spaces.

The class of problems we are interested in is characterized by the dependence of the main part of differential operator on the functional parameter. The shape and topology optimization requires the analysis of singular perturbations of domains of integrations. From numerical point of view in the spirit of the *experimental mathematics* we are interested in local solutions of optimality conditions for the problem under consideration. Since the problem is non-convex the first order optimality conditions are not sufficient to find the global solution. The differentiability of the cost functional for parametric optimization problems depends on the nature of such a problem. In the classical shape optimization the so-called Lie derivative of the shape functional is employed. This means that an admissible domain is homeomorphic to a given domain, in particular the simply connected domains are possible solutions if we start with a member of the class. In order to deal with the topology optimization problems we should admit the singular domain perturbations within the admissible class. Thus the shape derivative in the direction of a given vector field is no more possible to characterize the cost variation.

If a small hole or inclusion is introduced in the geometric domain of integration, the size of the hole is a small parameter which governs the singular perturbation of geometrical domain. The new concept which describes the variations of the cost is one-sided topological derivative. One-sided because the radius of the new hole in the geometrical domain is positive, that is, we cannot create any hole with the negative radius. This difficulty is overcome by the so-called homogenization method, however the homogenization method leads to generalized designs instead of geometrical domains, i.e., we loose the set of admissible domains replaced by the new set of generalized solutions given by the tensors of matrices of admissible coefficients of elliptic operators.

In order to obtain the necessary optimality conditions in shape and topology optimization for elliptic boundary value problems we construct an approximation of the solution in the perturbed domain. Using the approximate solution the topological derivative of the cost is evaluated. An optimal domain is characterized by the sign of the topological derivative. Thus, we obtain the free boundary problem for the unknown optimal topology. The solution approximation requires the local regularity of the solution to the state equation in the unperturbed domain in the vicinity of the centre of a hole. The regularity is necessary to employ the Taylor formula. The method of matched asymptotic expansions was introduced by Arlen Il'in for the Dirichlet problem. The second method is called compound asymptotic expansions and it was used in particular by Serguei Nazarov.

We are interested in numerical evaluation of unknown shapes or geometrical domains in the framework of shape optimization. It means that a shape functional is minimized within a given family of admissible shapes or domains. The functional depends upon the solution of a given state equation in the form of an elliptic boundary value problem. We are interested in evaluation of the shape functional for the small parameter using only the solutions of the state equation in the intact domain. This is possible for the elliptic boundary value problems by using e.g., the compound or matched asymptotic expansions. In this way the solution of the state equation is approximate and the expansion of the shape functional is derived. Once the result is known, we can propose some robust methods for identification of the first term of expansion which is called the topological derivative of the shape functional. Let us introduce the shape functional of the form

$$\Omega \to J(\Omega),$$

where the geometrical domain $\Omega \subset \mathbb{R}^d$ with $d = 1, 2, 3$ serves the integration domain of the state equations for most of applications in structural mechanics, inverse problems as well as control problems. The singularly perturbed geometrical domain Ω_ε is obtained as the difference $\Omega \setminus \overline{\omega_\varepsilon}$ where the small set $\varepsilon \to \omega_\varepsilon$ with the centre $\widehat{x} \in \Omega$ stands for the geometrical singular perturbation of Ω. Such a perturbation is called the topological variation of the domain of reference. We assume that for $\varepsilon \to 0$ the topological variation is reduced to the point $\widehat{x} \in \Omega$. To the topological variation of the domain we can associate one sided expansion of the shape functional. As a result of asymptotic analysis we obtain a function $\mathcal{T}(x)$, $x \in \Omega$, such that

$$J(\Omega_\varepsilon) = J(\Omega) + |\omega_\varepsilon| \mathcal{T}(\widehat{x}) + o(|\omega_\varepsilon|).$$

In order to obtain nontrivial results we assume that the expansion takes the form

$$J(\Omega_\varepsilon) = J(\Omega) + f(\varepsilon) \mathcal{T}(\widehat{x}) + o(f(\varepsilon)),$$

where the function $\varepsilon \to f(\varepsilon)$ is to be determined from the asymptotic approximation of the solutions to the state equation. The choice of the latter function is made to obtain the necessary optimality conditions for the topology optimization problem under study. It turns out that $f(\varepsilon) = |\omega_\varepsilon|$ e.g., for the homogeneous Neumann conditions on the hole ω_ε in the case of the second order elliptic state equation. The variation of the shape functional associated with the creation of a small singularity around $\widehat{x} \in \Omega$ is measured by the function

$$\widehat{x} \mapsto \mathscr{T}(\widehat{x}),$$

which is called the topological derivative of the mapping $\Omega \mapsto J(\Omega)$.

We can propose a strategy to evaluate the topological derivative. The domain decomposition technique is useful to this end. First, a small neighbourhood of the singularity is consider and the expansion of the energy functional is derived. Such an expansion provides us with the expansion of the boundary Dirichlet to Neumann operator for the neighbourhood. As a result the expansion of pseudodifferential Steklov–Poincaré operator is obtained for the truncated domain. Using this expansion the initial singular problem is transformed into a regular problem however only in the truncated domain. Using such a construction a method of evaluation of topological derivatives can be proposed in the truncated domain. Thus, the weak solutions of boundary value problem are approximated in the truncated domain by the solutions to the problem with the perturbed bilinear form. In the truncated domain the singularity located at the point $\widehat{x} \in \Omega$ is surrounded by a small circle or a small sphere usually denoted by Γ_R. The perturbation of bilinear form which lives on the circle or on the sphere is determined from the expansion of the Steklov–Poincaré operator.

Notation for the truncated domain. We briefly describe the notation in an abstract setting to be used for the construction of an approximation of the shape functional in the truncated domain. The singularly perturbed domain is denoted by Ω_ε, the intact domain is Ω. Let us consider the specific shape functional $\Omega \to I(u)$ where $u := u(\Omega) \in H(\Omega)$ is a weak solution of the elliptic problem in variational form

$$a(\Omega; u, v) = L(\Omega; v) \quad \text{for all} \quad v \in H(\Omega).$$

In the same way the perturbed shape functional is defined in Ω_ε with the solution of the state equation which lives in the perturbed domain, $\varepsilon \to u := u(\Omega_\varepsilon) \in H(\Omega_\varepsilon)$ given by

$$a(\Omega_\varepsilon; u, v) = L(\Omega_\varepsilon; v) \quad \text{for all} \quad v \in H(\Omega_\varepsilon).$$

We construct an approximation of the perturbed shape functional $\Omega_R \to I_\varepsilon(u_\varepsilon)$ for the small parameter $\varepsilon > 0$. The domain $\Omega_R \subset \Omega$ is the truncated domain with removed subset which contains the small singularity located inside of the ball $B_R(\widehat{x})$. Here $u_\varepsilon \in H(\Omega)$ solves the approximate boundary value problem $a_\varepsilon(u_\varepsilon, v) = L_\varepsilon(v)$ for all $v \in H(\Omega)$. The perturbed bilinear form $a_\varepsilon(u, v) = a(u, v) + b_\varepsilon(u, v)$ is defined in the intact domain Ω. The bilinear form $(u, v) \to b_\varepsilon(u, v)$ is defined from the expansion of the Steklov–Poincaré operator and it is supported on $\Gamma_R = \partial B_R(\widehat{x})$.

1.2 Elementary Example of Topological Derivative

Topological derivatives of shape functionals were introduced for elliptic BVPs quite recently. Instead of deformations of domains by diffeomorphism [83, 246], asymptotic analysis in singularly perturbed geometrical domains is performed for the purposes of shape sensitivity analysis [206]. However, the approach based on deformations applies to all types of linear PDEs. In addition, the velocity method of shape sensitivity analysis is simpler compared to asymptotic analysis, but it has some drawbacks from the point of view of numerics. In general, there is a close relationship between the two approaches; the results obtained by the second approach can also be derived from the velocity method, under some local regularity assumptions on solutions of elliptic BVPs. In other words, the knowledge of shape gradients and shape Hessians leads to topological derivatives of, e.g., energy functionals under additional regularity assumptions. Asymptotic analysis is performed in the original, unperturbed domains. It is worth mentioning that classical shape sensitivity analysis can be derived by using only asymptotic analysis tools. In the main body of this chapter, for simplicity, we restrict ourselves to the simple case of circular or ball-shaped voids. In order to fix ideas, let us give a very simple example.

Example 1.1 The notion of topological derivative extends the conventional definition of derivative to functionals depending on a *geometrical domain*, subjected to singular topology changes. The analogy between $\mathcal{T}(x)$ and the corresponding expressions for a conventional derivative should be noted. To illustrate the application of this concept, let us consider the very simple functional

$$J(\Omega) := |\Omega| = \int_{\Omega} 1, \tag{1.1}$$

with $\Omega \subset \mathbb{R}^2$ subject to the class of topological perturbations given by the nucleation of circular holes, namely $\omega_\varepsilon = B_\varepsilon(\widehat{x}) := \{\|x - \widehat{x}\| < \varepsilon\}$, for $\widehat{x} \in \Omega$. For two-dimensional domains Ω the functional $J(\Omega)$ represents the area of the domain. The expansion with respect to ε in this case is trivial:

$$J(\Omega_\varepsilon) = |\Omega_\varepsilon| = \int_{\Omega} 1 - \int_{B_\varepsilon} 1 = J(\Omega) - \pi\varepsilon^2. \tag{1.2}$$

The topological derivative $\mathcal{T}(x)$ and the function $f(\varepsilon)$ are immediately found to be

$$\mathcal{T}(x) = -1, \quad f(\varepsilon) = \pi\varepsilon^2. \tag{1.3}$$

In this particular case $\mathcal{T}(x)$ is independent of x, and the remainder term of the topological asymptotic expansion vanishes.

Remark 1.1 A particular case of shape optimization consists in minimizing a functional over an admissible set of characteristic functions [246]. In Example 1.1 the integral is of this type,

$$J(\Omega) = \int_{\mathbb{R}^2} \chi_\Omega(x)dx,$$

where χ_Ω is the characteristic function of Ω.

1.3 Shape and Topology Optimization

Shape and topology optimization is a broad field of modern research in pure (differential geometry) and applied mathematics, and in computational mechanics. In applied mathematics, it is a branch of the calculus of variations, partial differential equations and numerical methods. In computational mechanics, design and synthesis of metamaterials, fracture and damage modelling and structural optimization are areas of particular interest for shape and topology optimization techniques. Shape and topology optimization is also an efficient mathematical tool for numerical solution of inverse problems, involving defect identification for instance.

Mathematical analysis of shape optimization problems. The problem of shape optimization is to minimize a shape functional over a family of admissible domains. The shape functional depends directly on geometrical domains Ω and implicitly by means of solutions to the state equation $u = u(\Omega)$, defined in Ω. For example, in structural mechanics the specific functional depends on solutions to elasticity BVPs defined in the domains of integration:

$$\Omega \mapsto \mathcal{J}_\Omega(u).$$

From the mathematical point of view, the questions to address, as usual in the calculus of variations, are:

1. *Existence of an optimal domain such that*

$$\mathcal{J}_{\Omega^*}(u^*) \le \mathcal{J}_\Omega(u).$$

   The set of admissible domains is selected in such a way that the minimizing sequence admits a limit and the passage to the limit in the state equation can be performed.
2. *Necessary conditions for optimality*. In the case of differentiable shape functionals, the shape differentiability of solutions to the state equation is sufficient for the necessary optimality conditions under some regularity assumptions for the optimal domain. For singular perturbations of geometrical domains the optimality conditions are obtained with the topological derivatives. The latter technique can be used for the topology optimization problems.
3. *Convergence of numerical methods*. To our best knowledge, convergence of solution to the shape optimization problem is an open question in shape optimization.

Level set method with topological derivatives. The main idea of numerical methods based on topological derivatives is the construction of an auxiliary level set function, depending on the topological derivative in the reference domain [24, 25, 126, 129]. In this method the shape gradient is simply replaced by the topological derivative. The line search procedure defined by simple rules in terms of the descent direction, given by the current value of the topological derivative, is used to modify the current shape. The shape is determined by the level set function. Therefore, within the topological derivative method, there is no need for the complicated Hamilton-Jacobi equation to control the shape evolution. During the optimization procedure, the shape evolution is governed by a simple updating algorithm based on the current topological derivative. In numerical examples the procedure converges to a local solution of the topology-shape optimization problem. Therefore this method belongs to the field of experimental mathematics. To our best knowledge, the convergence of the level set method is still to be shown, except for some particular cases. First order methods are discussed in the book, with details and examples. In contrast to the first order method [25], a novel method based on second order topological derivatives has been recently introduced [67]. The two-term expansion of the functional is exploited, leading to a quadratic and strictly convex form with respect to the parameters under consideration. Therefore, a single optimization step gives the solution, without the need for an iterative algorithm. Second order methods are also discussed in details and examples are given.

Bibliography on the subject. There is a vast literature on the subject; the representative sources are, e.g., the monographs [61, 83, 124, 225, 246] on the theory and applications in mechanics of solids and fluids. In general we cannot expect the existence of a global solution to shape optimization problems, since shape functionals are nonconvex. Therefore, it makes sense to introduce generalized optimal solutions of shape optimization problems, e.g., by means of the homogenization technique [3, 46, 151]. As a result, the optimization procedure leads to optimal microstructures and there appear subdomains of optimal domains filled with metamaterials. These methods are known in structural mechanics as the homogenization method [47] or the SIMP method [1]. In this book we are interested in applications of asymptotic analysis tools and techniques to shape and topology optimization in singularly perturbed geometrical domains [200, 202, 203, 206]. In particular, the topological derivative concept can be seen as a particular case of the broader class of asymptotic methods fully developed in the books by Ammari and Kang [15], Ammari et al. [12] and Maz'ya, Nazarov and Plamenevskij [187, 188], for instance. Regarding the theoretical development of topological asymptotic analysis, see for instance [22, 32, 35, 98, 105, 145, 171, 198, 204, 210, 215, 219, 229, 242, 243]. An account of new developments in this branch of shape optimization may be found in the book by Novotny and Sokołowski [220].

1.4 Evaluation of Topological Derivatives

Evaluation of topological derivatives requires approximation of changes to solutions of elliptic BVPs with respect to small singular perturbations of geometrical domains. Such approximations are constructed, e.g., in the monographs [134, 188]; see also [201, 204]. We refer the reader to [171] for the comparison of the known methods for evaluation of energy change due to the appearance of cavities in elastic solids. We point out that in all methods listed below, for differentiable shape functionals appropriate adjoint state equations can be introduced. Evaluation methods for topological derivatives depend on specific applications. The computational approaches of particular interest are:

The direct method of shape calculus combined with asymptotic expansions of solutions. It is proposed in the first paper on the topic [239]. In this method the Taylor expansion of shape functionals, obtained in the framework of the velocity method [246], is used to pass to the limit with the small perturbation parameter, and as a result the topological derivative of the specific shape functional is obtained. The method is improved in [216, 241], and finally uses the first shape derivatives only for a class of shape functionals. The method is difficult to apply for variational inequalities. See, e.g., [38] for an application of standard asymptotic analysis to the Signorini problem, performed under the hypothesis of strict complementarity for unknown solutions. In the case of linear elliptic equations, the method can be combined with an adjoint equation approach [98].

The direct method of two-scale asymptotic analysis. It is performed in [206] for the elliptic systems. For the energy it results in the self-adjoint extensions of elliptic operators. In this method appropriate adjoint state equations can be introduced at the end of the procedure, in order to simplify formulas for topological derivatives. In other words, first a two-scale asymptotic approximation of solutions with prescribed precision is constructed, and it is then used to construct an approximation of the functional. Finally, a convenient form of topological derivative is obtained, for the purposes of numerical methods. In [206], complete proofs of the results obtained for linearized elasticity boundary value problems are given.

The method using fundamental solutions in truncated domains. It is combined with the standard two-scale expansion techniques of asymptotic analysis (see e.g., [231]). In a sense, this method is substantially improved by the addition of the domain decomposition technique with the Steklov–Poincaré operators, see the last point of this list. In addition, we refer to [22] for compound asymptotic analysis combined with a modified adjoint sensitivity method, which leads to a constructive approach for deriving the topological asymptotic expansion.

The method using the technique of integral equations. It is proposed for instance in [13, 121]. We also refer to the recent book [11] for modelling influence of defects in elastic bodies, using well-established asymptotic formulas, with some applications to imaging.

The domain decomposition technique. It is combined with the asymptotic expansions of Steklov–Poincaré operators for small defects [243]. This framework is well adapted to the sensitivity analysis of coupled models in multiphysics, as well as of variational inequalities and contact problems in elasticity. The adjoint state equations are introduced in order to simplify the formulas for topological derivatives. The method has been used in many numerical examples. See, e.g., [10, 107].

The evaluation technique depends on the problem under consideration. If there is a state equation, the evaluation process usually includes asymptotic expansions of solutions with respect to a small parameter $\varepsilon \to 0$, which governs the size of singular geometrical perturbation. We restrict ourselves to elliptic BVPs, namely, the Laplace or Helmholtz equations, as well as to the BVPs of linear elasticity or for the Stokes system. The expansion depends on the spatial dimensions $d = 2, 3$, since we use the fundamental solutions to the associated elliptic equations. Most important for applications are elliptic equations in three spatial dimensions.

The differentiability of energy functionals is the first task of asymptotic analysis for the purposes of topological optimization. This property of functionals is used explicitly in numerical methods, and implicitly when using the domain decomposition technique. The general mathematical theory of solution's expansions, which applies to the elasticity system, can be found e.g. in [206]; see also [214] for the polarization tensors associated with the elasticity system. The results are given for arbitrary shapes of cavities or holes. However, closed formulas are available only for some specific shapes [214]. Usually, the results obtained for the contrast parameter $0 < \gamma < \infty$ lead to the limit cases for rigid inclusions and cavities [220]. Once the result of asymptotic analysis is known, there are some methods available which can be used to identify topological derivatives.

We are particularly interested in numerical methods which are used in the framework of topology optimization. Therefore, topological derivatives should be given by robust expressions, which are approximated with the standard finite element techniques. The recommended way of identification is the application of the domain decomposition technique, in particular for topological derivatives of coupled models in multiphysics [243]. See also [231] for an early attempt at the truncated domain approach without Steklov–Poincaré operators. In order to simplify asymptotic analysis, we are going to explain the application of Steklov–Poincaré operators to evaluation of topological derivatives for elliptic BVPs. Without the state equation, this evaluation is simpler, it becomes a purely geometrical problem; we refer to [82] for recent results in this direction.

To obtain the form of topological derivatives, appropriate asymptotic analysis of solutions to the associated partial differential equation should be performed. There are some monographs on the subject, e.g., [134, 188]. We also refer the reader to the related results obtained in collaboration with Serguei A. Nazarov [73, 160, 206–212, 214]. See also [36] for further developments, using matched asymptotic expansions for the Laplacian. In general, the form of the topological derivative is given in terms of adjoint states and polarization tensors. This is an additional difficulty for numerical methods of topology optimization using topological derivatives, since the polarization tensor is known explicitly only for specific geometries of voids or inclusions.

1.5 Open Problems for Topological Derivative Method

The topological derivative method in shape and topology optimization introduces the asymptotic analysis of elliptic BVPs, e.g., into the field of structural optimization in elasticity. The method requires the local regularity of solutions to elliptic problems. Nowadays, classical shape optimization techniques are not restricted to elliptic problems, but can be applied to evolution problems, including linear parabolic and hyperbolic equations.

Extension of the topological derivative method to evolution problems is one of the challenging issues in the field of shape optimization. In particular, for transport equations, the notion of topological derivative is still to be discovered. Let us mention that the transport equations are components of compressible Navier–Stokes equations. The modern theory of shape optimization for compressible Navier–Stokes equations can be found in the monograph [225]. Another domain which is promising for developments of shape optimization is nonlinear elasticity. Evolution of geometrical domains is also used in growth modelling. Shape and topology optimization for nonlinear elasticity is still poorly understood. There is already some numerical evidence that topological derivatives are well adapted to the design of metamaterials in elasticity.

In order to state some open problems for applications of topological sensitivity analysis to numerical solution of shape optimization and inverse problems, we specify the mathematical framework which combines analysis of weak solutions to elliptic BVPs with the domain decomposition method, as well as with asymptotic approximation of solutions in singularly perturbed geometrical domains.

Applications of Γ-convergence to topological differentiability of shape functionals. It is known that the shape derivatives of energy type functionals can be obtained by the Γ-limit procedure; see e.g. the derivation of the elastic energy with respect to crack length [147, 148] or [57, 58, 144, 146, 168] for related topics. It is not known if the same Γ-convergence can be used for topological derivatives.

Sensitivity analysis of evolution variational inequalities. Shape optimization for stationary variational inequalities is studied in [246]. The results are based on Hadamard differentiability of the metric projection onto convex sets in Sobolev spaces [122, 195, 245]. The results are obtained by using potential theory in Dirichlet spaces [102], which leads to Hadamard differentiability of the metric projection. For stationary variational inequalities with local constraints on gradients or on stresses, shape and topological sensitivity analysis is an open problem. For example, for Hencky plasticity there are no results, either in shape optimization or for topological derivatives. Extension of such results to evolution variational inequalities is a challenging problem, necessary for development of topological derivatives.

Asymptotic analysis of evolution variational inequalities. Asymptotic analysis of stationary variational inequalities was studied e.g. by Argatov and Sokolowski [38]. The concept of polyhedral subsets of Sobolev spaces can be used in order to derive topological derivatives for contact problems in solid mechanics. Open problems for stationary variational inequalities include some models of plates or shells and plasticity. This domain of research has stagnated for a long time. The mathematical result required concerns directional differentiability of the metric projection onto a convex set defined by local constraints on the stresses. Extending such results to asymptotic analysis of evolution variational inequalities is another challenging problem.

Second order necessary optimality conditions in topology optimization. First order necessary optimality conditions are known for linear state equations [242]. Using second order topological derivatives to derive optimality conditions seems to be an open problem.

Exact solutions in singularly perturbed geometrical domains. Exact solutions of elliptic BVPs in a ring are used to evaluate topological derivatives within the domain decomposition technique. Such results are of interest for the PDEs listed below.

Elasticity system. Using complex potentials of Kolosov, one can obtain exact solutions of the elasticity system in two spatial dimensions. These results lead to topological derivatives of arbitrary order. Extending such results to a full range of models in mechanics is an interesting line of research, which is still to be taken up.

Wave equations. An important field of applications is electromagnetism and exact solutions, which are not standard from the mathematical point of view for asymptotic analysis. We refer the reader to [19, 43, 255] for related results in the static case. Open problems in the field of wave equations cover all aspects of time-dependent problems, including asymptotic analysis in singularly perturbed geometrical domains. Wave equations with self-adjoint extensions of elliptic operators are used in [156].

Transport equations. The domain decomposition method is also used for transport problems. Open problems in this field involve all aspects of asymptotic analysis. The compressible Navier–Stokes equations can also be considered from the point of view of singular domain perturbations, provided the related results for the transport equation component are established.

Nonlinear problems. The proposed approach to evaluation of topological derivatives applies to some classes of variational inequalities [178] and to coupled boundary value problems of multiphysics [10]. From the mathematical point of view, the domain decomposition method for evaluation of topological derivatives uses only the asymptotic expansions of energy functionals in a small neighbourhood of the singular domain perturbation, produced by a hole or a cavity. For our applications, the energy expansions give rise to the expansions of nonlocal Steklov–Poincaré boundary operators. As a result for the truncated domain, the perturbations of boundary conditions obtained with the expansion of Steklov–Poincaré boundary operators lead to regular perturbations of the bilinear form, and allow us to avoid the so-called self-adjoint extensions of elliptic operators in punctured domains. Topological derivatives are known for linear elliptic BVPs. They remain almost unknown for some nonlinear problems, including nonlinear elasticity and plasticity. See [26] for some recent developments in this direction. Extension of this approach to shape optimization of evolution problems should also be considered for variational inequalities modelling compressible fluids.

1.6 Description of the Content of the Book

In Chap. 2 the theory of asymptotic expansions of solutions to elliptic boundary value problems in singularly perturbed geometrical domains is briefly described for the purposes of topological derivative method. The Helmholtz problem is considered in Chap. 3. The known formulas for the topological derivatives usually contain the point-wise values of the solutions as well as of the gradients of the solutions for the boundary value problems in the intact or unperturbed geometrical domain. Such values are usually required at the centre of the hole or of the cavity. The numerical evaluation of point-wise values imposes some restrictions on the finite element method used for computations. Therefore, it is useful to obtain an equivalent form of the topological derivative which is written in terms of the solution evaluated in the intact domain and restricted to the boundary of the hole or the cavity. This is possible by an applications of the modified Bessels functions for the Helmholtz equation in two spatial variables. In Chap. 4, the topological derivative of the optimal value of the cost functional for a class of optimal control problems is obtained. The closed form of the topological derivative is given. Such an approach allows the simultaneous optimum design of a structure for the purposes of the optimal control. The first order necessary conditions for optimality are derived for a model problem in Chap. 5. Rough speaking, we have the sign of the topological derivative in a domain which satisfies the conditions.

A first order topology design algorithm based on the topological derivative concept combined with a level-set domain representation method is presented in Chap. 6, together with its applications in the context of topology optimization. The first order method described in Chap. 6 is then applied for solving different problems in engineering. In particular, Chap. 7 deals with synthesis of compliant thermomechanical actuators. A multiphysics problem in piezoelectricity is considered in Chap. 8. The device transforms electrical energy supplemented via its piezoceramic part into elastic energy. This feature is used in the synthesis of compliant piezomechanical actuators. Chapter 9 deals with unilateral contact problem in elasticity. The domain decomposition method developed in Chap. 2 is applied in order to obtain the associated topological derivative. A real-life case study concerning the redesign of an eyebar belonging to the eyebars chain of a cable bridge is presented.

A second order reconstruction algorithm of Newton-type based on higher order topological derivatives is proposed in Chap. 10, which is specific designed for solving a class of inverse reconstruction problems. Finally, in Chap. 11 the electrical impedance tomography (EIT) problem is considered, which consists in determining the distribution of the electrical conductivity of a medium subject to a set of current fluxes, from measurements of the corresponding electrical potentials on its boundary.

Chapter 2
Theory in Singularly Perturbed Geometrical Domains

The topological derivative method is used for numerical solution of shape and topology optimization problems. We consider singularly perturbed geometrical domains for elliptic boundary value problems. Therefore, we admit holes or cavities in the interior or singular perturbations of boundaries. In contrast to traditional approaches for shape and topology optimization, the topological derivative has been specifically designed to deal with shape and topology optimization. It has been introduced by Sokołowski and Żochowski in the fundamental paper [239] to fill a gap in the existing literature at that time. Actually, the idea was to give a precise (mathematical) answer to the following question: What does happen when a hole is nucleated? The answer to this question is not trivial at all. In fact, when a hole is nucleated, singularities would appear. Therefore, in order to deal with this problem, the theory of asymptotic analysis in singularly perturbed geometrical domain is need, which is developed in this chapter with help of the domain decomposition technique combined with the Steklov–Poincaré pseudodifferential boundary operator. As the main result, we show that the expansion of the energy coincides with the expansion of the Steklov–Poincaré operator on the boundary of the truncated domain, which allows for deriving an expression for the topological derivative.

2.1 Preliminaries

Topological derivatives of shape functionals are introduced for elliptic boundary value problems quite recently. Instead of deformations of domains by diffeomorphism [246], the asymptotic analysis in singularly perturbed geometrical domains is considered for the purposes of shape sensitivity analysis [206]. However, the first approach applies to all types of linear PDEs. In addition, the velocity method of shape sensitivity analysis is simpler compared to asymptotic analysis, but it has

A. A. Novotny et al., *Applications of the Topological Derivative Method*,
Studies in Systems, Decision and Control 188,
https://doi.org/10.1007/978-3-030-05432-8_2

some drawbacks from the point of view of numerical methods. In general, there is a close relationship between two types of approaches, the results obtained by the second approach can be also derived from the velocity method under some regularity assumptions for the elliptic boundary value problems. In another words, the knowledge of shape gradients and shape Hessians leads to the topological derivatives of e.g., the energy functionals under additional regularity assumptions. The asymptotic analysis is performed in the intact, unperturbed domains and locally requires the appropriate regularity of solutions. It is worth to say, that the classical shape sensitivity analysis can be performed by using the asymptotic analysis tools, only. We speak on the ideas for elliptic problems and the singular perturbations of the principal part of the elliptic operator. There are also many problems when only the lower order terms of the operator are perturbed, and in such a case asymptotic analysis substantially simplifies. It is crucial for applications to know the exact form of topological derivatives. Sometimes, the obtained expressions for topological derivatives depend on the unknown polarization tensor for the material which is used to build the geometrical domains. Therefore, we present the method of topological derivative combined with the domain decomposition technique. The topological derivatives are evaluated in a small subdomain for the energy functional. Then the asymptotic expansion of the energy is propagated to the truncated domain via the boundary Steklov–Poincaré operator. Thus, the topological derivative is obtained for an arbitrary shape functional defined on the truncated domain.

2.2 Asymptotic Expansions for the Domain Decomposition Technique

The domain decomposition technique seems to be the most important method of evaluation of topological derivatives for numerical methods of topology optimization. In particular, the topological derivatives for BVPs of coupled models in multiphysics can be obtained in the framework of this technique, combined with asymptotic expansions of Steklov–Poincaré operator. In control theory, at least for scalar elliptic problems, the Steklov–Poincaré operator becomes the Dirichlet-to-Neumann map. The domain decomposition method can be applied to linear elliptic problems, as well as to variational inequalities. Numerical applications of the method can be found e.g. for shape-topology optimization in piezo-elasticity [10], or in thermo-elasticity [107].

2.2.1 *Asymptotic Expansions of Steklov–Poincaré Operators*

Our goal is to apply the domain decomposition method to evaluate topological derivatives. The first step of such an evaluation for complex models is always the

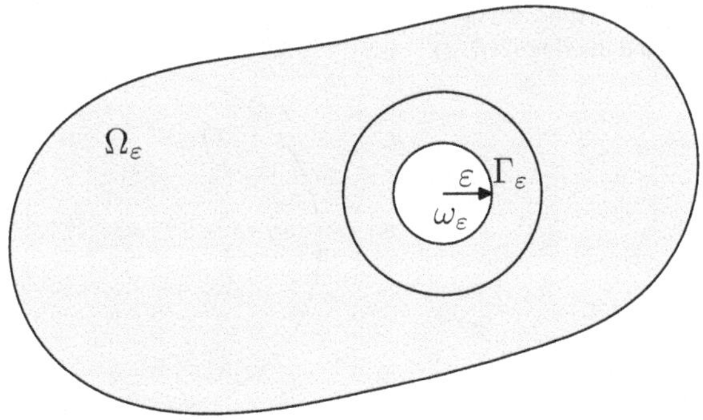

Fig. 2.1 Domain Ω_ε with small void

local analysis of singular perturbations of geometrical domains. Thus, e.g., in linear elasticity in two or three spatial dimensions, with a traction free hole or cavity, we consider a ring-shaped domain $C(R, \varepsilon)$ and obtain an asymptotic expansion of the Steklov–Poincaré operators associated to the elasticity problem, defined on its external boundary Γ_R, with respect to the small parameter $\varepsilon \to 0$.

Let us consider a family of perturbations Ω_ε of the reference domain Ω by small holes or small cavities $\omega_\varepsilon(\widehat{x})$, with center $\widehat{x} \in \Omega$. See Fig. 2.1. The method consists in approximating singular domain perturbations by regular perturbations of the bilinear form $v \mapsto a(\Omega; v, v)$, in the variational formulation of the elliptic boundary value problem. The approximation means that the small domain ω_ε is replaced by the correction term to the bilinear form, given by the boundary bilinear form $v \mapsto \varepsilon^d b(\Gamma_R; v, v)$ concentrated on the curve or surface Γ_R. This bilinear form can be determined from asymptotic expansions of Steklov–Poincaré operators defined on the interface Γ_R, i.e., from the topological derivative of the energy functional in $C(R, \varepsilon)$ for small $R > \varepsilon > 0$. In this section we provide all the necessary details with some examples.

In a sense our approach is similar, but it is not equivalent, to the so-called self-adjoint extensions of elliptic operators [223] as they are used in physics. See e.g. [206, 207, 209, 210] for applications to asymptotic approximations, which lead to equivalent formulas for topological derivatives. In other words, we are able to define an approximate mathematical model in the original domain in such a way that the first order asymptotic expansion of the energy functional is the same as for the original model in the perturbed domain. For self-adjoint extensions the domain Ω_ε is replaced by the punctured domain $\Omega \setminus \{\widehat{x}\}$, while in our approach the truncated domain is $\Omega_R := \Omega \setminus \overline{B_R(\widehat{x})}$. See Fig. 2.2. This approximation is sufficient for most of the applications we have in mind. In any case we need polarization tensors or matrices [73, 214] in order to use topological derivatives in numerical methods applied to shape and topology optimization.

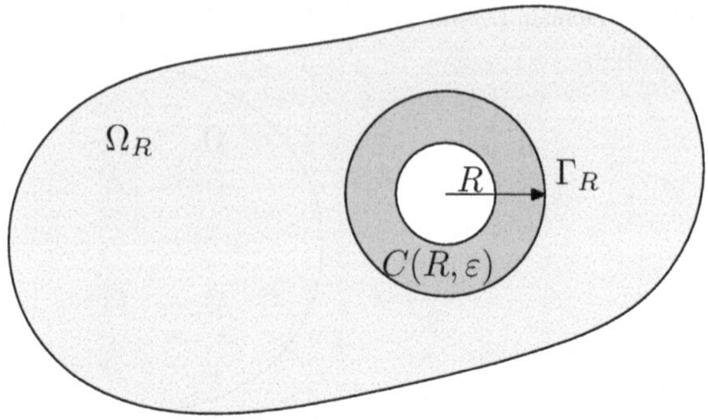

Fig. 2.2 Truncated domain Ω_R and the ring $C(R, \varepsilon)$

2.2.2 *From Singular Domain Perturbations to Regular Perturbations of Bilinear Forms in Truncated Domains*

The main idea of evaluation of topological derivatives by an application of the domain decomposition technique is the asymptotic analysis of the so-called Steklov–Poincaré boundary operators. The strategy for analysis is divided into two steps. First, the singular domain perturbations are considered in a simple geometry for the energy functionals (see exception for the Helmholtz problem). Then, the regular perturbations of the bilinear form are defined in the truncated domain.

In this section we present the abstract framework of asymptotic analysis for solutions of variational problems, posed in singularly perturbed geometrical domains. For simplicity let us consider linear problems. A weak solution of a linear elliptic problem with a symmetric, coercive and continuous bilinear form posed in $\Omega \subset \mathbb{R}^d$,

$$u := u_\Omega \in H \; : \; a(\Omega; u, \varphi) = L(\Omega; \varphi) \qquad \forall \varphi \in H, \tag{2.1}$$

is given by a unique minimizer of the quadratic functional

$$I(\Omega; \varphi) := \frac{1}{2} a(\Omega; \varphi, \varphi) - L(\Omega; \varphi) = \frac{1}{2} a(\varphi, \varphi) - L(\varphi) \tag{2.2}$$

over the Sobolev space $H := H(\Omega)$ of functions defined on Ω. For simplicity we write also

$$I(\varphi) := \frac{1}{2} a(\Omega; \varphi, \varphi) - L(\Omega; \varphi).$$

The energy shape functional is defined for Ω as

$$\Omega \mapsto \mathscr{E}(\Omega) = I(u) = \frac{1}{2} a(\Omega; u, u) - L(\Omega; u).$$

We consider the singular geometrical perturbation $\Omega_\varepsilon := \Omega \setminus \overline{B_\varepsilon(\widehat{x})}$ of the reference domain produced by a small circular hole or cavity.

In order to evaluate the topological derivatives of the energy shape functional, as well as of some other shape functionals, we are going to use the domain decomposition technique. To this end the reference domain is divided into two subdomains. The complement in Ω of the first subdomain $\Omega_R := \Omega \setminus \overline{B_R(\widehat{x})}$ is the ball $B_R(\widehat{x})$ which includes the singular geometrical perturbation of the reference domain. The energy functional in the perturbed domain is

$$\mathscr{E}(\Omega_\varepsilon) = \mathscr{E}(\Omega_R) + \mathscr{E}(C(R,\varepsilon)),$$

where $C(R,\varepsilon)$ is a ring,

$$C(R,\varepsilon) := B_R(\widehat{x}) \setminus \overline{B_\varepsilon(\widehat{x})}.$$

Now, we would like to introduce a boundary perturbation of the bilinear form

$$(\varphi,\varphi) \mapsto b_\varepsilon(\Gamma_R;\varphi,\varphi)$$

such that for $\varepsilon > 0$,

$$\mathscr{E}(C(R,\varepsilon)) = b_\varepsilon(\Gamma_R; u_\varepsilon, u_\varepsilon).$$

In fact, we can introduce such a form which is asymptotically exact for the first order expansion of the solution u_ε, restricted to the truncated domain Ω_R, namely

$$u_\varepsilon^R = u^R + \varepsilon^d q^R + o(\varepsilon^d) \tag{2.3}$$

in $H(\Omega_R)$ for fixed $R > \varepsilon > 0$. Here $f(\varepsilon) = \varepsilon^d$, thus we assume implicitly that the homogeneous Neumann boundary conditions are prescribed on the cavity $B_\varepsilon(\widehat{x})$. In this way we could obtain the first order topological derivatives for shape functionals defined in Ω_R, using the expansion of the energy functional in the ring $C(R,\varepsilon)$.

Let us fix $\widehat{x} \in \Omega$ and consider again the truncated domain $\Omega_R := \Omega \setminus \overline{B_R(\widehat{x})}$ and the perturbed domain $\Omega_\varepsilon := \Omega \setminus \overline{B_\varepsilon(\widehat{x})}$, where $R > \varepsilon > 0$ are two parameters such that $\varepsilon \to 0$ and $\overline{B_\varepsilon(\widehat{x})} \subset \Omega$. By definition (2.2) we associate with the domains the quadratic functionals

$$I_R(\varphi) := \frac{1}{2}a(\Omega_R;\varphi,\varphi) - L(\Omega_R;\varphi) = \frac{1}{2}a_R(\varphi,\varphi) - L_R(\varphi)$$

and

$$I_\varepsilon(\varphi) := \frac{1}{2}a_\varepsilon(\varphi,\varphi) - L_\varepsilon(\varphi), \tag{2.4}$$

obtained by restriction of the test functions $\varphi \in H(\Omega)$ to Ω_R (respectively to Ω_ε). Our goal is to construct an approximation of the quadratic functional for which the minimizer u_ε^R is given by the restriction to Ω_R of the variational solution u_ε in the singularly perturbed domain. The variational solution in Ω is given by (2.1) and the variational problem in the perturbed domain Ω_ε is given by

$$u_\varepsilon \in H_\varepsilon \ : \ a_\varepsilon(u_\varepsilon, \varphi) = L_\varepsilon(\varphi) \qquad \forall \varphi \in H_\varepsilon . \tag{2.5}$$

To this end we introduce the nonlocal Steklov–Poincaré operator $\mathscr{A}_\varepsilon$ on the interior boundary Γ_R of Ω_R. The operator is defined by the nonhomogeneous Dirichlet boundary value problem over $C(R, \varepsilon)$. We determine the expansion of the Steklov–Poincaré operator

$$\mathscr{A}_\varepsilon = \mathscr{A} + \varepsilon^d \mathscr{B} + \mathscr{R}_\varepsilon$$

in the space of linear operators, and introduce the bilinear form associated with the first term of the expansion,

$$b(h, h) := (\mathscr{B}(h), h)_{\Gamma_R}.$$

It can be shown that minimization of the first order approximation of the quadratic functional (2.6), defined in the original domain, leads to the first order expansion of the minimizers for the perturbed domain, which holds in the truncated domain Ω_R:

Theorem 2.1 *Let u_ε^R be the minimizer of the approximate quadratic functional*

$$I_\varepsilon^R(\varphi) = \frac{1}{2} a(\Omega; \varphi, \varphi) + \frac{1}{2} \varepsilon^d (\mathscr{B}(\varphi), \varphi)_{\Gamma_R} - L(\Omega_R; \varphi). \tag{2.6}$$

Then the restriction of this minimizer to Ω_R (denoted by the same symbol u_ε^R) has in $H^1(\Omega_R)$ an expansion

$$u_\varepsilon^R = u^R + \varepsilon^d q^R + o(\varepsilon^d), \tag{2.7}$$

where $u^R = u_{|\Omega_R}$. This expansion coincides with the expansion of the solution to (2.5) in $H^1(\Omega_R)$.

Corollary 2.1 *In the case of variational equations resulting from the minimization of (2.4) we have the same result. Indeed, the first order expansion of the minimizers $\varepsilon \mapsto u_\varepsilon$, restricted to the truncated domain Ω_R, i.e., for $R > \varepsilon > 0$,*

$$u_{\varepsilon|\Omega_R} = u_{|\Omega_R} + \varepsilon^d q^R + o(\varepsilon^d) \tag{2.8}$$

is preserved when using the minimization of (2.6), since, e.g., for the second order elliptic boundary value problems,

$$\|u_{\varepsilon|\Omega_R} - u_\varepsilon^R\|_{H^1(\Omega_R)} = o(\varepsilon^d).$$

Taking into account these estimates, the topological derivatives of shape functionals, defined by integrals in the truncated domain, can be obtained.

Corollary 2.2 *The topological derivative of the tracking type functional,*

$$J(\Omega_\varepsilon) = \frac{1}{2}\int_{\Omega_R} (u_\varepsilon - z_d)^2 dx, \tag{2.9}$$

is given by the expression

$$\mathcal{T}(\widehat{x}) = \int_{\Omega_R} (u^R(x) - z_d(x))q^R(x)dx = \int_{\Omega_R} (u(x) - z_d(x))q(x)dx. \tag{2.10}$$

where

$$q = \lim_{\varepsilon\to 0} \frac{1}{\varepsilon^d}(u_\varepsilon - u) \quad in \quad \Omega_R.$$

Remark 2.1 The topological derivative of the tracking type functional (2.9) can be simplified for variational equations by using the adjoint state equation. Then the full potential of the topological derivative is seen, since it becomes a function of the point $\widehat{x}$, obtained at the expense of solving only two sets of state-like systems. However, even if we cannot use adjoint equations, the domain decomposition method based on the asymptotic expansion of the Steklov–Poincaré operator has its advantages. They consist in replacing the influence of the small hole (inclusion) by the small, localized disturbance of the main bilinear form (2.6), defined on the original domain. In this way, the method facilitates and speeds up numerical approximations, e.g. using finite element methods.

2.2.3 Signorini Problem in Two Spatial Dimensions

Now we shall explain the domain decomposition technique used in approximation of quadratic energy functionals to evaluate topological derivatives for the Laplace operator [243]. We consider the Signorini problem in a domain $\Omega \subset \mathbb{R}^2$ with smooth boundary $\partial\Omega = \Gamma_0 \cup \Gamma_s$. Let us restrict ourselves to the homogeneous Neumann boundary conditions on holes in order to use $f(\varepsilon) \approx \varepsilon^2$. The bilinear form

$$a(u, v) = a(\Omega; u, v) = \int_\Omega \nabla u \cdot \nabla v dx$$

is coercive and continuous over the Sobolev space

$$H^1_{\Gamma_0}(\Omega) = \{v \in H^1(\Omega) \mid v = 0 \text{ on } \Gamma_0\},$$

and the linear form

$$L(v) = L(\Omega; v) = \int_\Omega f v dx$$

is continuous on $L^2(\Omega)$. There is a unique solution to the variational inequality

$$u \in K \ : \ a(u, v-u) \geq L(v-u) \qquad \forall v \in K,$$

where K is the convex and closed cone

$$K = K(\Omega) = \{v \in H^1_{\Gamma_0}(\Omega) \mid v \geq 0 \text{ on } \Gamma_s\}.$$

Let us consider the variational inequality over the singularly perturbed domain $\Omega_\varepsilon = \Omega \setminus \overline{B}_\varepsilon$,

$$u_\varepsilon \in K_\varepsilon = K(\Omega_\varepsilon) \ : \ a_\varepsilon(u_\varepsilon, v-u_\varepsilon) \geq L_\varepsilon(v-u_\varepsilon) \qquad \forall v \in K_\varepsilon,$$

with the solution given by the unique minimizer of the quadratic functional

$$I_\varepsilon(v) = \frac{1}{2}a_\varepsilon(v,v) - L_\varepsilon(v) \tag{2.11}$$

over the convex set $K_\varepsilon := K(\Omega_\varepsilon)$. Let us assume that L_ε is supported in Ω_R.

We can show that there is an approximation of (2.11), denoted by

$$I^R_\varepsilon(v) = \frac{1}{2}a_\varepsilon(\Omega_R; v,v) - L_\varepsilon(\Omega_R; v),$$

such that the first order expansion of the minimizers with respect to the small parameter $\varepsilon^2 \to 0$ is the same in $H^1(\Omega_R)$ as in the perturbed problem. Namely, we can introduce a continuous, symmetric and nonlocal bilinear form on the circle $\Gamma_R = \{\|x - \widehat{x}\| = R > \varepsilon > 0\}$,

$$H^{1/2}(\Gamma_R) \times H^{1/2}(\Gamma_R) \ni (v,v) \mapsto b(\Gamma_R; v,v) \in \mathbb{R},$$

such that

$$I^R_\varepsilon(v) = \frac{1}{2}a(\Omega; v,v) - L(\Omega; v) + \varepsilon^2 b(\Gamma_R; v,v).$$

Furthermore, if the solution to the perturbed variational inequality has an expansion in $H^1(\Omega_R)$,

$$u_\varepsilon = u + \varepsilon^2 q + o(\varepsilon^2),$$

then the minimizer u^R_ε of $v \mapsto I^R_\varepsilon(v)$ over the convex cone $K(\Omega)$ has in $H^1(\Omega_R)$ the same first order expansion,

$$u^R_\varepsilon = u + \varepsilon^2 q + o(\varepsilon^2).$$

Corollary 2.3 *The topological derivative of the tracking type functional for the variational inequality,*

$$J(\Omega_\varepsilon) = \frac{1}{2}\int_{\Omega_R} (u_\varepsilon - z_d)^2 dx, \tag{2.12}$$

is given by

$$\mathcal{T}(\widehat{x}) = \int_{\Omega_R} (u^R(x) - z_d(x))q^R(x)dx = \int_{\Omega_R} (u(x) - z_d(x))q(x)dx. \tag{2.13}$$

where

$$q = \lim_{\varepsilon\to 0} \frac{1}{\varepsilon^2}(u_\varepsilon - u) \quad in \ \Omega_R.$$

For variational inequalities we cannot introduce adjoint states. However, we are able to replace the singular geometrical domain perturbation $B_\varepsilon(\widehat{x})$ by the regular perturbation $v \mapsto \varepsilon^2 b(\Gamma_R; v, v)$ of the bilinear form $v \mapsto a(\Omega; v, v)$, and preserve the first order expansion of minimizers in the truncated domain. The bilinear form is constructed using the expansion of the Steklov–Poincaré operator, defined on Γ_R and given by the expansion of the energy functional in the ring $C(R, \varepsilon)$:

$$a(\Omega_\varepsilon; v, v) \cong a(\Omega; v, v) + \varepsilon^2 b(\Gamma_R; v, v),$$

where

$$b(\Gamma_R; v, v) = \left(\frac{1}{\pi R^3}\int_{\Gamma_R} v\cdot(x_1 - \widehat{x}_1)\, ds\right)^2 + \left(\frac{1}{\pi R^3}\int_{\Gamma_R} v\cdot(x_2 - \widehat{x}_2)\, ds\right)^2.$$

2.2.4 Domain Decomposition Method for Elasticity

The domain decomposition method has been used to evaluate topological derivatives for coupled BVPs [10, 107, 220]. In this section we shall consider asymptotic corrections to the energy functional corresponding to the elasticity system in $\mathbb{R}^d$, where $d = 2, 3$. The change of energy is caused by creating a small ball-like void of variable radius ε in the interior of the domain Ω, with the homogeneous Neumann boundary condition on its surface. We assume that the void is centred at the origin $\mathcal{O}$. We take $\Omega_R = \Omega \setminus \overline{B_R}$, where $B_R := B(\mathcal{O}, R)$ is an open ball with fixed radius R. In this way the void $B_\varepsilon := B(\mathcal{O}, \varepsilon)$ is surrounded by $B_R \subset \Omega$. We also denote the ring or spherical shell as $C(R, \varepsilon) = B_R \setminus \overline{B_\varepsilon}$, and its boundaries as $\Gamma_R = \partial B_R$ and $\Gamma_\varepsilon := \partial B_\varepsilon$.

Using these notations we define our main tool, the Dirichlet-to-Neumann mapping for linear elasticity, which is called the Steklov–Poincaré operator:

$$\mathscr{A}_\varepsilon : H^{1/2}(\Gamma_R; \mathbb{R}^d) \to H^{-1/2}(\Gamma_R; \mathbb{R}^d)$$

by means of the boundary value problem

$$\begin{aligned} &\mu \Delta w + (\lambda + \mu)\nabla(\operatorname{div} w) = 0, \quad \text{in} \ \ C(R, \varepsilon), \\ &w = v \ \ \text{on} \ \ \Gamma_R, \quad \sigma(u)n = 0 \ \ \text{on} \ \ \Gamma_\varepsilon, \end{aligned} \tag{2.14}$$

so that

$$\mathscr{A}_\varepsilon(v) = \sigma(w)n \ \ \text{on} \ \ \Gamma_R. \tag{2.15}$$

Here, μ, λ are the Lamé coefficients, and $\sigma(w)$ is the Cauchy stress tensor,

$$\sigma(w) = 2\mu(\nabla w)^s + \lambda \operatorname{div}(w)\, \mathrm{I}, \quad \text{with} \quad (\nabla w)^s = \frac{1}{2}(\nabla w + (\nabla w)^\top).$$

Let u^R be the restriction of u to Ω_R, and $\gamma^R(\varphi)$ the trace of φ on Γ_R, still denoted by φ for simplicity. We may then define the functional

$$I_\varepsilon^R(\varphi) = \frac{1}{2}\int_{\Omega_R} \sigma(\varphi) \cdot (\nabla\varphi)^s\, dx - \int_{\Gamma_N} h \cdot \varphi\, ds + \frac{1}{2}\int_{\Gamma_R} \mathscr{A}_\varepsilon(\varphi) \cdot \varphi\, ds \tag{2.16}$$

and the solution u_ε^R as a minimizer for

$$I_\varepsilon^R(u_\varepsilon^R) = \inf_{\varphi \in K \subset H(\Omega_R)} I_\varepsilon^R(\varphi), \tag{2.17}$$

We have replaced the variable domain $\varepsilon \mapsto \Omega_\varepsilon$ by a fixed truncated domain Ω_R, at the price of introducing a variable boundary operator $\mathscr{A}_\varepsilon$. Thus, the goal is to find an asymptotic expansion

$$\mathscr{A}_\varepsilon = \mathscr{A} + \varepsilon^d \mathscr{B} + \mathscr{R}_\varepsilon, \tag{2.18}$$

where the remainder $\mathscr{R}_\varepsilon$ is of order $o(\varepsilon^d)$ in the operator norm in

$$\mathscr{L}(H^{1/2}(\Gamma_R; \mathbb{R}^2), H^{-1/2}(\Gamma_R; \mathbb{R}^2))$$

and the operator $\mathscr{B}$ is regular enough, namely it is bounded and linear:

$$\mathscr{B} \in \mathscr{L}(L^2(\Gamma_R; \mathbb{R}^2), L^2(\Gamma_R; \mathbb{R}^2)).$$

Under this assumption the following propositions hold true.

Proposition 2.1 *Assume that* (2.18) *holds in the operator norm. Then*

$$u_\varepsilon^R \to u^R \tag{2.19}$$

strongly in the $H^1(\Omega_R)$*-norm.*

Proposition 2.2 *The energy functional has a representation*

$$I_\varepsilon^R(u_\varepsilon^R) = I(u) + \varepsilon^d \langle \mathscr{B}(u), u\rangle_R + o(\varepsilon^d)\,, \tag{2.20}$$

where $o(\varepsilon^d)/\varepsilon^d \to 0$ *as* $\varepsilon \to 0$ *in the same energy norm.*

Here $I^R(u^R)$ denotes the functional I_ε^R on the original domain, i.e. $\varepsilon = 0$, with $\mathscr{A}_\varepsilon$ replaced by $\mathscr{A}$ applied to truncation of u. Generally, the energy correction for the elasticity system has the form

$$\langle \mathscr{B}(u^R), u^R\rangle_R \;=\; -c_d e_u(\mathscr{O}),$$

where $c_d = \mathrm{vol}(B_1)$, with B_1 being the unit ball in $\mathbb{R}^d$. The energy-like density function $e_u(\mathscr{O})$ has the form [220, 242]

$$e_u(\mathscr{O}) = \frac{1}{2}\mathbb{P}\sigma(u^R)\cdot(\nabla u^R)^s(\mathscr{O}),$$

where for $d = 2$ and the plane stress,

$$\mathbb{P} \;=\; \frac{1}{1-\nu}(4\mathbb{I} - \mathrm{I}\otimes\mathrm{I}),$$

and for $d = 3$,

$$\mathbb{P} \;=\; \frac{1-\nu}{7-5\nu}\left(10\mathbb{I} - \frac{1-5\nu}{1-2\nu}\mathrm{I}\otimes\mathrm{I}\right).$$

Here $\mathbb{I}$ is the fourth order identity tensor, and I is the second order identity tensor.

This approach is important for variational inequalities, since it allows us to derive formulas for topological derivatives which are similar to expressions obtained for the corresponding linear BVPs.

Explicit form of the operator $\mathscr{B}$ in two spatial dimensions. Let us denote, for the plane stress case,

$$k = \frac{\lambda+\mu}{\lambda+3\mu}.$$

It has been proved in [243] that the following exact formulas hold:

$$u_{1,1}(\mathscr{O}) + u_{2,2}(\mathscr{O}) = \frac{1}{\pi R^3}\int_{\Gamma_R}(u_1x_1 + u_2x_2)\,ds,$$

$$u_{1,1}(\mathscr{O}) - u_{2,2}(\mathscr{O}) = \frac{1}{\pi R^3}\int_{\Gamma_R}\left[(1-9k)(u_1x_1 - u_2x_2) + \frac{12k}{R^2}(u_1x_1^3 - u_2x_2^3)\right]ds,$$

$$u_{1,2}(\mathscr{O}) + u_{2,1}(\mathscr{O}) = \frac{1}{\pi R^3}\int_{\Gamma_R}\left[(1+9k)(u_1x_2 + u_2x_1) - \frac{12k}{R^2}(u_1x_2^3 + u_2x_1^3)\right]ds.$$

These expressions contain additional integrals of third powers of x_i. Therefore, strains evaluated at $\mathcal{O}$ may be expressed as linear combinations of integrals over the circle of the form

$$\int_{\Gamma_R} u_i x_j \, ds, \qquad \int_{\Gamma_R} u_i x_j^3 \, ds.$$

The same is true, due to Hooke's law, for the stresses $\sigma_{ij}(\mathcal{O})$. They may then be inserted into the expression for $\mathcal{B}$, yielding

$$\langle \mathcal{B}(u^R), u^R \rangle_R = -\frac{1}{2} c_2 \mathbb{P}\sigma(u) \cdot (\nabla u)^s.$$

These formulas are quite easy to compute numerically.

Explicit form of the operator $\mathcal{B}$ in three spatial dimensions. It turns out that a similar situation holds in three spatial dimensions, but obtaining exact formulas is more difficult. Assuming given values of u on Γ_R, the solution of the elasticity system in B_R may be expressed, following partially the derivation from [180, pp. 285 ff.], as

$$u = \sum_{n=0}^{\infty} [U_n + (R^2 - r^2) k_n(\nu) \nabla (\mathrm{div} U_n)]. \tag{2.21}$$

where $k_n(\nu) = 1/2[(3-2\nu)n - 2(1-\nu)]$ and $r = \|x\|$, with ν denoting the Poisson ratio. In addition,

$$U_n = \frac{1}{R^n} \left[a_{n0} d_n(x) + \sum_{m=1}^{n} (a_{nm} c_n^m(x) + b_{nm} s_n^m(x)) \right]. \tag{2.22}$$

The vectors

$$a_{n0} = (a_{n0}^1, a_{n0}^2, a_{n0}^3)^\top,$$

$$a_{nm} = (a_{nm}^1, a_{nm}^2, a_{nm}^3)^\top,$$

$$b_{nm} = (b_{nm}^1, b_{nm}^2, b_{nm}^3)^\top$$

are constant and the set of functions

$$\{d_0;\ d_1, c_1^1, s_1^1;\ d_2, c_2^1, s_2^1, c_2^2, s_2^2;\ d_3, c_3^1, s_3^1, c_3^2, s_3^2, c_3^3, s_3^3;\ \ldots\}$$

constitutes a complete system of orthonormal harmonic polynomials on Γ_R, related to Laplace spherical functions. Specifically,

$$c_k^l(x) = \frac{\hat{P}_k^{l,c}(x)}{\|\hat{P}_k^{l,c}\|_R}, \quad s_k^l(x) = \frac{\hat{P}_k^{l,s}(x)}{\|\hat{P}_k^{l,s}\|_R}, \quad d_k = \frac{P_k(x)}{\|\hat{P}_k\|_R}.$$

For example,

$$c_3^2(\mathbf{x}) = \frac{1}{R^4}\sqrt{\frac{7}{240\pi}}(15x_1^2x_3 - 15x_2^2x_3),$$

If the value of u on Γ_R is assumed to be given, then, denoting

$$\langle\phi, \psi\rangle_R = \int_{\Gamma_R} \phi\psi \, ds,$$

we have for $n \geq 0$, $m = 1..n$, $i = 1, 2, 3$:

$$\begin{aligned} a_{n0}^i &= R^n \langle u_i, d_n(x)\rangle_R, \\ a_{nm}^i &= R^n \langle u_i, c_n^m(x)\rangle_R, \\ b_{nm}^i &= R^n \langle u_i, s_n^m(x)\rangle_R. \end{aligned} \tag{2.23}$$

Since we are looking for $u_{i,j}(\mathcal{O})$, only the part of u which is linear in x is relevant. It contains two terms:

$$\hat{u} = U_1 + R^2 k_3(\nu)\nabla(\mathrm{div}U_3). \tag{2.24}$$

For any $f(x)$, $\nabla\mathrm{div}(af) = H(f)\cdot a$, where a is a constant vector and $H(f)$ is the Hessian matrix of f. Therefore

$$\begin{aligned} \hat{u} = &\frac{1}{R}[a_{10}d_1(x) + a_{11}c_1^1(x) + b_{11}s_1^1(x))] \\ &+ R^2k_3(\nu)\frac{1}{R^3}\Big[H(d_3)(x)a_{30} + \sum_{m=1}^{3}\big(H(c_3^m)(x)a_{3m} + H(s_3^m)(x)b_{3m}\big)\Big] \end{aligned} \tag{2.25}$$

From the above we may single out the coefficients of x_1, x_2, x_3 in u_1, u_2, u_3. For example,

$$\begin{aligned} u_{1,1}(\mathcal{O}) = &\frac{1}{R^3}\sqrt{\frac{3}{4\pi}}a_{11}^1 + \frac{1}{R^5}k_3(\nu)\Bigg[-3\sqrt{\frac{7}{4\pi}}a_{30}^3 9\sqrt{\frac{7}{24\pi}}a_{31}^1 \\ &-3\sqrt{\frac{7}{24\pi}}b_{31}^2 + 30\sqrt{\frac{7}{240\pi}}a_{32}^3 + 90\sqrt{\frac{7}{1440\pi}}a_{33}^1 + 90\sqrt{\frac{7}{1440\pi}}b_{33}^2\Bigg], \\ u_{1,2}(\mathcal{O}) = &\frac{1}{R^3}\sqrt{\frac{3}{4\pi}}(b_{11}^1 + a_{11}^2) + \frac{1}{R^5}k_3(\nu)\Big[-3\sqrt{\frac{7}{24\pi}}a_{31}^2 - \sqrt{\frac{7}{24\pi}}b_{31}^1 \\ &+ 15\sqrt{\frac{7}{60\pi}}b_{32}^3 - 90\sqrt{\frac{7}{1440\pi}}a_{33}^2 + 90\sqrt{\frac{7}{1440\pi}}b_{33}^1\Big]. \end{aligned}$$

Observe that

$$u_{1,1}(\mathcal{O}) + u_{2,2}(\mathcal{O}) + u_{3,3}(\mathcal{O}) = \frac{1}{R^3}\sqrt{\frac{3}{4\pi}}\big(R\langle u_1, c_1^1\rangle_R + R\langle u_2, s_1^1\rangle_R + R\langle u_3, d_1\rangle_R\big)$$

and $c_1^1 = \frac{1}{R^2}\sqrt{\frac{3}{4\pi}}x_1$, $s_1^1 = \frac{1}{R^2}\sqrt{\frac{3}{4\pi}}x_2$, $d_1 = \frac{1}{R^2}\sqrt{\frac{3}{4\pi}}x_3$, which is in agreement with the fact that the trace of the strain tensor is a harmonic function. As a result, the operator $\mathscr{B}$ may be defined by

$$\langle \mathscr{B}u^R, u^R\rangle_R = -c_3\mathbb{P}\sigma(u)\cdot(\nabla u)^s(\mathcal{O}),$$

where the right-hand side consists of integrals of u multiplied by first and third order polynomials in x_i over Γ_R resulting from (2.23). This is a very similar situation to the case of two spatial dimensions. Thus, the new expressions for strains make it possible to rewrite $\mathscr{B}$ in the form of the desired regularity. We refer the reader e.g. to [18, 50, 233] for supplementary material on elasticity models.

2.3 Matched Asymptotic Expansions for Neumann Problem

The natural question is to identify the general method which can be applied to obtain the form of the topological derivative for elliptic problems. There are two methods of matched or compound asymptotic expansions. We describe in details one of the methods for the Neumann problem in three spatial dimensions. For the convenience of the reader the asymptotic analysis of nonhomogeneous boundary value problem is performed for a simple model problem. The small cavity $\omega_\varepsilon := \varepsilon\omega$ with the centre at the origin $\mathcal{O} \in \omega_\varepsilon \subset \omega$ can be considered without loss of generality. In general we denote by the same symbol $\omega_\varepsilon(\widehat{x}) := \widehat{x} + \varepsilon\omega$ the cavity with the centre at $\widehat{x} \in \Omega$. The matched asymptotic expansions are used in two spatial dimensions for scalar problems with the Laplacian.

2.3.1 Asymptotic Expansion of the Steklov–Poincaré

We denote the smooth domain $\Omega_\varepsilon := \Omega \setminus \overline{\omega_\varepsilon}$ for $\varepsilon \to 0$ and let us consider the nonhomogeneous Dirichlet problem with $h \in H^{1/2}(\Gamma)$,

$$\begin{cases} \Delta w_\varepsilon = 0 \text{ in } \Omega_\varepsilon\ , \\ w_\varepsilon = h \text{ on } \Gamma\ , \\ \partial_n w_\varepsilon = 0 \text{ on } \partial\omega_\varepsilon\ , \end{cases} \tag{2.26}$$

where $\Gamma = \partial\Omega$ is the boundary of the intact domain Ω. The energy associated with (2.26) is given by the symmetric bilinear form on fractional Sobolev space $H^{1/2}(\Gamma)$

$$a_\varepsilon(h,h) = \int_{\Omega_\varepsilon} \|\nabla w_\varepsilon\|^2 dx.$$

We are interested in the asymptotic expansion of the quadratic functional for $\varepsilon \to 0$. To this end the technique of matched asymptotic expansions [134, 187, 188] is used.

Using the Green's formula, we derive the equivalent forms of the energy, here the boundary integrals stand for the duality pairing between the space $H^{1/2}(\Gamma)$ and its dual $H^{-1/2}(\Gamma)$,

$$a_\varepsilon(h,h) = \int_\Gamma w_\varepsilon \frac{\partial w_\varepsilon}{\partial n} ds = \int_\Gamma h \frac{\partial w_\varepsilon}{\partial n} ds \tag{2.27}$$

We use the method of matched asymptotic expansions, and look for two types of expansions, the outer expansion valid far from the cavity ω_ε

$$w_\varepsilon(x) = w_0(x) + \varepsilon^2 w_1(x) + \varepsilon^3 w_2(x) + \dots$$

and the inner expansion, valid in a small neighborhood of ω_ε

$$w_\varepsilon(x) = W_0(\xi) + \varepsilon W_1(\xi) + \varepsilon^2 W(\xi) + \dots,$$

where the fast variable ξ is defined by

$$\xi = \frac{x}{\varepsilon}.$$

Following [134, 187, 188] we obtain

$$W_0(\xi) \equiv w_0(0),$$

and

$$W_1(\xi) = \sum_{j=1}^{2} \mathcal{Y}^j(\xi) \frac{\partial w_0}{\partial x_j}(0),$$

where $\mathcal{Y}^j$ is harmonic in $\mathbb{R}^2 \setminus \overline{\omega}$ and $\omega := \omega_1$. In addition $\mathcal{Y}^j$ satisfies the homogeneous Neumann boundary conditions on $\partial\omega$ and enjoys the following behavior at infinity

$$\mathcal{Y}^j(\xi) = \xi_j + \frac{1}{2\pi\|\xi\|^2} \sum_{k=1}^{2} m^\omega_{kj} \xi_k + O(\|\xi\|^{-2}), \quad \|\xi\| \to \infty.$$

Its regular part is denoted by

$$\mathscr{Y}_0^j(\xi) := \frac{1}{2\pi\,\|\xi\|^2}\sum_{k=1}^{2} m_{kj}^{\omega}\xi_k + O(\|\xi\|^{-2}),$$

and we denote its higher order term $O(\|\xi\|^{-2})$

$$\tilde{\mathscr{Y}}_0^j(\xi) := \mathscr{Y}_0^j(\xi) - \frac{1}{2\pi\,\|\xi\|^2}\sum_{k=1}^{2} m_{kj}^{\omega}\xi_k,$$

Taking into account this expansion, we get

$$w_1(x) = \sum_{j,k=1}^{2}\frac{\partial w_0}{\partial x_j}(0)m_{kj}^{\omega}\mathscr{G}^{(k)}(x).$$

We denote by $\mathscr{G}^{(k)}$ the singular solutions to the problem posed in punctured domain

$$\begin{cases} \Delta_x\mathscr{G}^{(k)}(x) = 0 & \text{in } \Omega\setminus\mathscr{O}, \\ \mathscr{G}^{(k)}(x) = 0 & \text{on } \Gamma, \\ \mathscr{G}^{(k)}(x) = \dfrac{x_k}{2\pi\,\|x\|^2} + O(1) & \|x\|\to 0. \end{cases} \tag{2.28}$$

We put

$$\mathscr{G}^{(k)}(x) = \frac{x_k}{2\pi\,\|x\|^2} + \mathscr{G}_0^{(k)}(x),$$

where $\mathscr{G}_0^{(k)}$ stands for the regular part. Therefore, far from the cavity ω_ε, we have

$$w_\varepsilon(x) = w_0(x) + \varepsilon^2\sum_{j,k=1}^{2}\frac{\partial w_0}{\partial x_j}(0)m_{kj}^{\omega}\mathscr{G}^{(k)}(x) + O(\varepsilon^2).$$

Substituting this representation into formula (2.27) we obtain one term expansion

$$a_\varepsilon(h,h) = a(h,h) + \varepsilon^2 b(h,h) + O(\varepsilon^{3-\alpha}).$$

Here $\alpha\in(0,1)$ and

$$b(h,h) = \int_\Gamma h(x)\sum_{j,k=1}^{2}\frac{\partial w_0}{\partial x_j}(0)m_{kj}^{\omega}\frac{\partial\mathscr{G}^{(k)}}{\partial n}(x)ds$$

If we combine this with the integral equality on the sphere of radius $\delta > 0$

$$\int_{\mathbb{S}_\delta(\mathscr{O})}\left(x_j\frac{\partial}{\partial n}\frac{x_k}{2\pi\,\|x\|^2} - \frac{x_k}{2\pi\,\|x\|^2}\frac{\partial x_j}{\partial n}\right)ds = \delta_{jk}$$

we get

$$b(h,h) = -m^{\omega}\nabla w_0(0)\cdot\nabla w_0(0).$$

Since

$$\int_{\Gamma} w_0(x)\partial_n\mathscr{G}^{(k)}(x)ds = -\frac{\partial w_0}{\partial x_k}(0),$$

it follows that

$$b(h,h) = -\left(\int_{\Gamma} h(x)\partial_n\mathscr{G}^{(j)}(x)ds\right)m^{\omega}_{jk}\left(\int_{\Gamma} h(x)\partial_n\mathscr{G}^{(k)}(x)ds\right)ds.$$

Remark 2.2 It can be shown that the following supremum taken with respect to $H^{1/2}(\Gamma)$-norm is bounded with respect to $\varepsilon\to 0$,

$$\sup_{\|h\|\le 1}\left|a_\varepsilon(h,h)-a(h,h)-\varepsilon^2 b(h,h)\right|\le C_\alpha\varepsilon^{3-\alpha}.$$

Since the operators associated to bilinear forms $(h,h)\mapsto a_\varepsilon(h,h)$ are positive and self-adjoint, the one term expansion of Steklov–Poincaré operators is obtained for $\varepsilon\to 0$,

$$\mathscr{A}_\varepsilon = \mathscr{A} - \varepsilon^2\mathscr{B} + O(\varepsilon^{3-\alpha})$$

with the remainder bounded in the operator norm $H^{1/2}(\Gamma)\mapsto H^{-1/2}(\Gamma)$. The self-adjoint positive linear operators $\mathscr{A}_\varepsilon$ are uniquely determined by the symmetric and coercive bilinear forms $h\mapsto a_\varepsilon(h,h)$. The operator $\mathscr{B}$ is determined by $h\mapsto b(h,h)$.

2.3.2 Asymptotic Expansion of the Linear Form

Let us now consider the linear form

$$L_\varepsilon(h) = \int_{\Omega_\varepsilon} f(x)w_\varepsilon(x)dx$$

We use the method of matched asymptotic expansions and set

$$w_\varepsilon(w) = w_0(x) + \varepsilon\sum_{j=1}^{2}\frac{\partial w_0}{\partial x_j}(0)\mathscr{Y}_0^j(\xi) + \varepsilon^2\sum_{j,k=1}^{2}\frac{\partial w_0}{\partial x_j}(0)m^{\omega}_{jk}\mathscr{G}_0^{(k)}(x) + \dots$$

hence

$$L_\varepsilon(h) = \int_{\Omega_\varepsilon} f(x)w_0(x)dx + \varepsilon \int_{\Omega_\varepsilon} f(x) \sum_{j=1}^{2} \frac{\partial w_0}{\partial x_j}(0)\mathscr{Y}_0^j(\xi)$$

$$+ \varepsilon^2 \int_{\Omega_\varepsilon} f(x) \sum_{j,k=1}^{2} \frac{\partial w_0}{\partial x_j}(0) m_{jk}^\omega \mathscr{G}_0^{(k)}(x)dx.$$

Taking into account that

$$\left|\mathscr{Y}_0^j(\xi)\right| \le C_0 \frac{1}{\|\xi\|} = C_0 \frac{\varepsilon}{\|x\|} \quad \text{in} \quad \Omega_\varepsilon$$

it follows that

$$L_\varepsilon(h) = \int_{\Omega} f(x)w_0(x)dx - \int_{\omega_\varepsilon} f(x)w_0(x)dx$$

$$+ \varepsilon^2 \int_{\Omega} f(x) \sum_{j,k=1}^{2} \frac{\partial w_0}{\partial x_j}(0) m_{jk}^\omega \mathscr{G}^{(k)}(x)dx + \dots$$

In order to replace the integrals over Ω_ε by the integrals over intact domain Ω we use the estimates

$$\varepsilon \int_{\omega_\varepsilon} f(x) \sum_{j,k=1}^{2} \frac{\partial w_0}{\partial x_j}(0) m_{jk}^\omega \frac{\varepsilon x_k}{2\pi \|x\|^2} dx \le C_0 \varepsilon^2 \sup_{x \in \overline{\Omega}} |f(x)| \int_0^\varepsilon \frac{1}{r} r dr,$$

$$\int_{\omega_\varepsilon} f(x) \sum_{j,k=1}^{2} \frac{\partial w_0}{\partial x_j}(0) m_{jk}^\omega \mathscr{G}_0^{(k)}(x) \le C_0 \varepsilon^2.$$

Finally,

$$L_\varepsilon(h) = L_0(h) - \varepsilon^2 f(0)w_0(0)|\omega| + \varepsilon^2 \int_{\Omega} f(x) \sum_{j,k=1}^{2} \frac{\partial w_0}{\partial x_j}(0) m_{jk}^\omega \mathscr{G}^{(k)}(x)dx + \dots$$

2.3.3 Asymptotic Expansion of the Energy Functional

The energy functional

$$\mathscr{J}_{\Omega_\varepsilon}(u_\varepsilon) = \frac{1}{2}\int_{\Omega_\varepsilon} \|\nabla u_\varepsilon\|^2 dx - \int_{\Omega_\varepsilon} f u_\varepsilon dx = \frac{1}{2}\int_{\Gamma} u_\varepsilon \frac{\partial u_\varepsilon}{\partial n} ds - \frac{1}{2}\int_{\Omega_\varepsilon} f u_\varepsilon dx$$

depends on solutions to the boundary value problem

$$\begin{cases} -\Delta u_\varepsilon = f \quad \text{in } \Omega_\varepsilon \,, \\ u_\varepsilon = h_\varepsilon \text{ on } \Gamma \,, \\ \partial_n u_\varepsilon = 0 \quad \text{on } \partial\omega_\varepsilon \,. \end{cases} \tag{2.29}$$

with the associated Green's formula

$$\int_{\Omega_\varepsilon} \|\nabla u_\varepsilon\|^2 dx = \int_\Gamma h_\varepsilon \frac{\partial u_\varepsilon}{\partial n} ds + \int_{\Omega_\varepsilon} f u_\varepsilon dx$$

Here, we assume that the Dirichlet boundary datum also depends on the small parameter

$$h_\varepsilon = h_0 + \varepsilon^2 h_1 + o(\varepsilon^2) \quad \text{in} \quad H^{1/2}(\Gamma)$$

and that there is a source term inside of the perturbated domain Ω_ε.

The approximation of solutions takes the form

$$u_\varepsilon(x) = v_0(x) + \varepsilon^2 v_1(x) + \varepsilon \sum_{j=1}^{2} \frac{\partial v_0}{\partial x_j}(0) \mathscr{Y}_0^j(\xi) + \varepsilon^2 \sum_{j,k=1}^{2} \frac{\partial v_0}{\partial x_j}(0) m_{jk}^\omega \mathscr{G}_0^{(k)}(x) + \ldots$$

where

$$\begin{cases} -\Delta v_0 = f \quad \text{in } \Omega \,, \\ v_0 = h_0 \text{ on } \Gamma \,, \end{cases} \quad \text{and} \quad \begin{cases} -\Delta v_1 = 0 \quad \text{in } \Omega \,, \\ v_1 = h_1 \text{ on } \Gamma \,. \end{cases}$$

The approximation of the normal derivatives

$$\frac{\partial u_\varepsilon}{\partial n}(x) = \frac{\partial v_0}{\partial n}(x) + \varepsilon^2 \frac{\partial v_1}{\partial n}(x) + \varepsilon \sum_{j=1}^{2} \frac{\partial v_0}{\partial x_j}(0) \frac{\partial \mathscr{Y}_0^j}{\partial \nu}(\xi) + \varepsilon^2 \sum_{j,k=1}^{2} \frac{\partial v_0}{\partial x_j}(0) m_{jk}^\omega \frac{\partial \mathscr{G}_0^{(k)}}{\partial n}(x) + \ldots$$

where $\dfrac{\partial}{\partial n} = n \cdot \nabla_x$, $\dfrac{\partial}{\partial \nu} = n \cdot \nabla_\xi$ and $\xi = x/\varepsilon$. We recall that the higher order term of $\mathscr{Y}_0^j(\xi)$ satisfies

$$\left|\tilde{\mathscr{Y}}_0^j(\xi)\right| \le C_0 \frac{1}{\|\xi\|^2} = C_0 \frac{\varepsilon^2}{\|x\|^2} \quad \text{for} \quad \xi = \frac{x}{\varepsilon} \in \mathbb{R}^2 \setminus \omega \quad \text{or for} \quad x \in \Omega_\varepsilon .$$

Thus, in the approximation of $u_\varepsilon(x)$ the terms of order $O(\varepsilon^3)$

$$\varepsilon \sum_{j=1}^{2} \frac{\partial v_0}{\partial x_j}(0) \tilde{\mathscr{Y}}_0^j(\xi)$$

can be neglected. Therefore, from the formula

$$u_\varepsilon(x) = v_0(x) + \varepsilon^2 v_1(x) + \varepsilon \sum_{j=1}^{2} \frac{\partial v_0}{\partial x_j}(0)\tilde{\mathscr{Y}}_0^j(\xi) + \varepsilon^2 \sum_{j,k=1}^{2} \frac{\partial v_0}{\partial x_j}(0) m_{jk}^\omega \mathscr{G}^{(k)}(x) + \dots$$

we deduce

$$\frac{\partial u_\varepsilon}{\partial n}(x) = \frac{\partial v_0}{\partial n}(x) + \varepsilon^2 \frac{\partial v_1}{\partial n}(x) + \varepsilon^2 \sum_{j,k=1}^{2} \frac{\partial v_0}{\partial x_j}(0) m_{jk}^\omega \frac{\partial \mathscr{G}^{(k)}}{\partial n}(x) + \dots$$

and it follows that

$$\begin{aligned} u_\varepsilon(x)\frac{\partial u_\varepsilon}{\partial n}(x) &= v_0(x)\frac{\partial v_0}{\partial n}(x) \\ &+ \varepsilon^2 \left(v_1(x)\frac{\partial v_0}{\partial n}(x) + v_0(x)\frac{\partial v_1}{\partial n}(x) + v_0(x) \sum_{j,k=1}^{2} \frac{\partial v_0}{\partial x_j}(0) m_{jk}^\omega \mathscr{G}^{(k)}(x) \right) + \dots \end{aligned}$$

since the second order term

$$\frac{\partial v_0}{\partial n}(x) \sum_{j,k=1}^{2} \frac{\partial v_0}{\partial x_j}(0) m_{jk}^\omega \mathscr{G}^{(k)}(x)$$

vanishes taking into account that $\mathscr{G}^{(k)}(x) = 0$ on the boundary Γ. We return to the shape functional

$$\mathscr{J}_{\Omega_\varepsilon}(u_\varepsilon) = \frac{1}{2}\int_\Gamma u_\varepsilon(x)\frac{\partial u_\varepsilon}{\partial n}(x) ds - \frac{1}{2}\int_{\Omega_\varepsilon} f u_\varepsilon dx$$

and find the approximations for the integrals,

$$\begin{aligned} \frac{1}{2}\int_\Gamma u_\varepsilon(x)\frac{\partial u_\varepsilon}{\partial n}(x) ds &= \frac{1}{2}\int_\Gamma v_0(x)\frac{\partial v_0}{\partial n}(x) ds \\ &+ \frac{\varepsilon^2}{2}\int_\Gamma \left(v_1(x)\frac{\partial v_0}{\partial n}(x) + v_0(x)\frac{\partial v_1}{\partial n}(x)\right) ds \\ &+ \frac{\varepsilon^2}{2}\int_\Gamma v_0(x) \sum_{j,k=1}^{2} \frac{\partial v_0}{\partial x_j}(0) m_{jk}^\omega \frac{\partial \mathscr{G}^{(k)}}{\partial n}(x) ds + \dots \end{aligned}$$

and

$$\begin{aligned} -\frac{1}{2}\int_{\Omega_\varepsilon} f u_\varepsilon dx &= -\frac{1}{2}\int_{\Omega_\varepsilon} f v_0 dx - \frac{\varepsilon^2}{2}\int_{\Omega_\varepsilon} f v_1 dx \\ &- \frac{\varepsilon^2}{2}\int_{\Omega_\varepsilon} f \sum_{j,k=1}^{2} \frac{\partial v_0}{\partial x_j}(0) m_{jk}^\omega \mathscr{G}^{(k)}(x) + \dots, \end{aligned}$$

which can be written as

$$-\frac{1}{2}\int_{\Omega_\varepsilon} fu_\varepsilon dx = -\frac{1}{2}\int_{\Omega} fv_0 dx + \frac{1}{2}\int_{\omega_\varepsilon} fv_0 dx - \frac{\varepsilon^2}{2}\int_{\Omega} fv_1 dx + \frac{\varepsilon^2}{2}\int_{\omega_\varepsilon} fv_1 dx$$
$$-\frac{\varepsilon^2}{2}\int_{\Omega} f\sum_{j,k=1}^{2}\frac{\partial v_0}{\partial x_j}(0)m^{\omega}_{jk}\mathscr{G}^{(k)}(x)dx + \frac{\varepsilon^2}{2}\int_{\omega_\varepsilon} f\sum_{j,k=1}^{2}\frac{\partial v_0}{\partial x_j}(0)m^{\omega}_{jk}\mathscr{G}^{(k)}(x)dx + \ldots$$

or as follows

$$-\frac{1}{2}\int_{\Omega_\varepsilon} fu_\varepsilon dx = -\frac{1}{2}\int_{\Omega} fv_0 dx + \frac{\varepsilon^2}{2} f(0)v_0(0)|\omega| - \frac{\varepsilon^2}{2}\int_{\Omega} fv_1 dx$$
$$-\frac{\varepsilon^2}{2}\int_{\Omega} f\sum_{j,k=1}^{2}\frac{\partial v_0}{\partial x_j}(0)m^{\omega}_{jk}\mathscr{G}^{(k)}(x)dx + O(\varepsilon^3)$$

Where we take into account that using the Taylor formula $\int_{\omega_\varepsilon} fv_0 dx$ is replaced by $\varepsilon^2 f(0)v_0(0)|\omega|$, in the same way it follows that $\varepsilon^2\int_{\omega_\varepsilon} fv_1 dx$ is $O(\varepsilon^4)$; finally, the latter integral over ω_ε is of the type $O\left(\int_0^\varepsilon \frac{1}{r} r\right)$. As a result

$$\mathscr{J}_{\Omega_\varepsilon}(u_\varepsilon) = \frac{1}{2}\int_{\Gamma} v_0(x)\frac{\partial v_0}{\partial n}(x)ds - \frac{1}{2}\int_{\Omega} fv_0 dx$$
$$+\frac{\varepsilon^2}{2}\int_{\Gamma}\left(v_1(x)\frac{\partial v_0}{\partial n}(x) + v_0(x)\frac{\partial v_1}{\partial n}(x)\right)ds + \frac{\varepsilon^2}{2}\int_{\Gamma} v_0(x)\sum_{j,k=1}^{2}\frac{\partial v_0}{\partial x_j}(0)m^{\omega}_{jk}\frac{\partial\mathscr{G}^{(k)}}{\partial n}(x)ds$$
$$-\frac{\varepsilon^2}{2}\int_{\Omega} f\sum_{j,k=1}^{2}\frac{\partial v_0}{\partial x_j}(0)m^{\omega}_{jk}\mathscr{G}^{(k)}(x)dx + \frac{\varepsilon^2}{2} f(0)v_0(0)|\omega| - \frac{\varepsilon^2}{2}\int_{\Omega} fv_1 dx.$$

We denote by $B_\delta(\mathscr{O})$ the ball at origin of radius δ, with its boundary $\mathbb{S}_\delta := \mathbb{S}_\delta(\mathscr{O})$. By the Green's formula in the domain $\Omega_\delta = \Omega \setminus \overline{B_\delta(\mathscr{O})}$, with the boundary $\partial\Omega_\delta = \Gamma \cup \mathbb{S}_\delta$, for $\delta \to 0$,

$$\int_{\Omega_\delta}(v_0\Delta\mathscr{G}^{(k)} - \mathscr{G}^{(k)}\Delta v_0)dx = \int_{\partial\Omega_\delta}\left(v_0\frac{\partial\mathscr{G}^{(k)}}{\partial n} - \mathscr{G}^{(k)}\frac{\partial v_0}{\partial n}\right)ds.$$

Since $\Delta v_0 = -f$, we find

$$\int_{\Omega_\delta}\mathscr{G}^{(k)} f dx = \int_{\Gamma} v_0\frac{\partial\mathscr{G}^{(k)}}{\partial n}ds + \int_{\mathbb{S}_\delta}\left(v_0\frac{\partial\mathscr{G}^{(k)}}{\partial n} - \mathscr{G}^{(k)}\frac{\partial v_0}{\partial n}\right)ds.$$

Passage to the limit $\delta \to 0$ leads to

$$\int_{\Omega_\delta}\mathscr{G}^{(k)} f dx - \int_{\Gamma} v_0\frac{\partial\mathscr{G}^{(k)}}{\partial n}ds = \frac{\partial v_0}{\partial x_k}(0)$$

Finally, we arrive at the expression

$$\mathscr{J}_{\Omega_\varepsilon}(u_\varepsilon) = \mathscr{J}_\Omega(v_0) + \frac{\varepsilon^2}{2} f(0)v_0(0)|\omega| - \frac{\varepsilon^2}{2} \sum_{j,k=1}^{2} \frac{\partial v_0}{\partial x_j}(0) m_{jk}^{\omega} \frac{\partial v_0}{\partial x_k}(0)$$
$$+ \frac{\varepsilon^2}{2} \int_\Gamma \left(v_1(x) \frac{\partial v_0}{\partial n}(x) + v_0(x) \frac{\partial v_1}{\partial n}(x) \right) ds - \frac{\varepsilon^2}{2} \int_\Omega f(x) v_1(x) dx + O(\varepsilon^3).$$

2.4 Asymptotics of Steklov–Poincaré Operators in Multilayer Subdomains

In this section the similar asymptotic analysis of elliptic boundary value problems in subdomain $\Omega_R \subset \mathbb{R}^2$ is performed, but we modify the situation, assuming that the hole is filled only partially, different material constituting a fixed part of it. In this way, we may consider double asymptotic expansions and the associated limit passage, where both the size of the hole, as well as the proportion of the different material contained in it can vary. From mechanical point of view this model describes e.g. to the hole with hardened walls. The analysis is based again on exact representation of solutions and allows to obtain the perturbation of solutions, using the fact that these solutions may be considered as minimizers of energy functional. The method is also suitable for double asymptotic expansions of solutions as well as energy form. The ultimate goal is to use obtained formulas in the evaluation of topological derivatives for elliptic boundary value problems.

2.4.1 *Multilayer Inclusions*

Let us consider the domain Ω containing the hole with boundary made of modified material as depicted in Fig. 2.3. For simplicity the hole is located at the origin of the coordinate system. In order to write down the model problem, we introduce some notations.

$$B_s = \{x \in \mathbb{R}^2 \; : \; \|x\| < s\}$$
$$C_{s,t} = \{x \in \mathbb{R}^2 \; : \; s < \|x\| < t\}$$
$$\Gamma_s = \{x \in \mathbb{R}^2 \; : \; \|x\| = s\}$$
$$\Omega_s = \Omega \setminus B_s$$

Then the problem in the intact domain Ω has the form

$$\begin{cases} k_1 \Delta w_0 = 0 & \text{in } \Omega, \\ w_0 = g_0 & \text{on } \partial\Omega. \end{cases} \tag{2.30}$$

The model problem in the modified domain reads:

$$\begin{cases} k_1 \Delta w_\varepsilon = 0 & \text{in } \Omega_\varepsilon, \\ w_\varepsilon = g_0 & \text{on } \partial\Omega, \\ w_\varepsilon = v_\varepsilon & \text{on } \Gamma_\varepsilon, \\ k_2 \Delta v_\varepsilon = 0 & \text{in } C_{\lambda\varepsilon,\varepsilon}. \end{cases} \tag{2.31}$$

$$\begin{cases} k_2 \partial_{n_2} v_\varepsilon = 0 \text{ on } \Gamma_{\lambda\varepsilon}, \\ k_1 \partial_{n_1} w_\varepsilon + k_2 \partial_{n_2} v_\varepsilon = 0 \text{ on } \Gamma_\varepsilon, \end{cases}$$

where n_1 and n_2 are exterior normal vectors to Ω_ε and to $C_{\lambda\varepsilon,\varepsilon}$, respectively, and $0 < \lambda < 1$.

We want to investigate the influence of the small ring-like inclusion made of another material on the difference $w_\varepsilon - w_0$ in Ω_R, where Γ_R surrounds $C_{\lambda\varepsilon,\varepsilon}$ and R is fixed. We assume that $\varepsilon \to 0+$ and λ is considered temporarily constant. If we define

$$u_\varepsilon = \begin{cases} w_\varepsilon & \text{in } \Omega_\varepsilon \\ v_\varepsilon & \text{in } C_{\lambda\varepsilon,\varepsilon} \end{cases}$$

then problem (2.31) reduces itself to finding the minimum of the energy functional

$$\mathscr{E}_1(u_\varepsilon) = \frac{1}{2}\int_{\Omega_\varepsilon} k_1 \nabla u_\varepsilon \cdot \nabla u_\varepsilon \, dx + \frac{1}{2}\int_{C_{\lambda\varepsilon,\varepsilon}} k_2 \nabla u_\varepsilon \cdot \nabla u_\varepsilon \, dx \tag{2.32}$$

for $u_\varepsilon \in H^1(\Omega_\varepsilon)$, $u_\varepsilon = g_0$ on $\partial\Omega$. This expression may be rewritten as

$$\mathscr{E}_1(u_\varepsilon) = \frac{1}{2}\int_{\Omega_R} k_1 \nabla w_\varepsilon \cdot \nabla w_\varepsilon \, dx + \frac{1}{2}\int_{C_{\varepsilon,R}} k_1 \nabla w_\varepsilon \cdot \nabla w_\varepsilon \, dx + \frac{1}{2}\int_{C_{\lambda\varepsilon,\varepsilon}} k_2 \nabla v_\varepsilon \cdot \nabla v_\varepsilon \, dx.$$

Using integration by parts we obtain

$$\mathscr{E}_1(u_\varepsilon) = \frac{1}{2}\int_{\Omega_R} k_1 \nabla w_\varepsilon \cdot \nabla w_\varepsilon \, dx + \frac{1}{2}\int_{\Gamma_\varepsilon} \left(w_\varepsilon k_1 \frac{\partial w_\varepsilon}{\partial n_1} + v_\varepsilon k_2 \frac{\partial v_\varepsilon}{\partial n_2} \right) ds + \frac{1}{2}\int_{\Gamma_R} k_1 w_\varepsilon \frac{\partial w_\varepsilon}{\partial n_3} \, ds,$$

where n_3 – exterior normal to Ω_R. Hence, due to boundary and transmission condition,

$$\mathscr{E}_1(u_\varepsilon) = \frac{1}{2}\int_{\Omega_R} k_1 \nabla w_\varepsilon \cdot \nabla w_\varepsilon \, dx + \frac{1}{2}\int_{\Gamma_R} k_1 w_\varepsilon \frac{\partial w_\varepsilon}{\partial n_3} \, ds \tag{2.33}$$

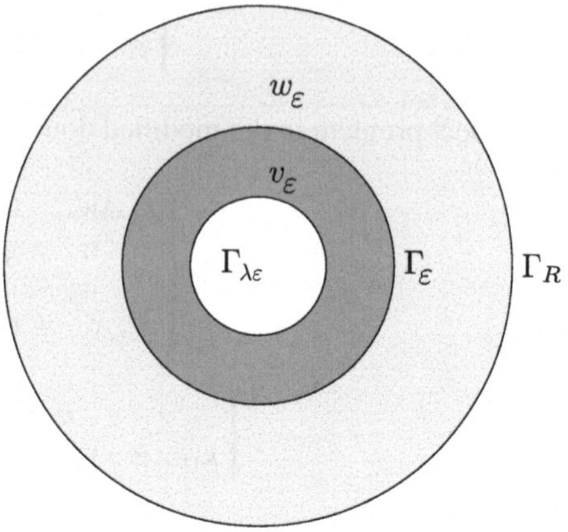

Fig. 2.3 Domains of definition for w_ε and v_ε

2.4.2 Steklov–Poincaré Operator in Multilayer Inclusion

Observe that $\mathscr{E}_1(w_0)$ corresponds to the problem (2.30). Therefore the main goal is to find the Steklov–Poincaré operator

$$\mathscr{A}_{\lambda,\varepsilon} \;:\; w \in H^{1/2}(\Gamma_R) \longmapsto \partial_{n_3} w_\varepsilon \in H^{-1/2}(\Gamma_R) \tag{2.34}$$

where the normal derivative is computed from auxiliary problem

$$\begin{cases} k_1 \Delta w_\varepsilon = & 0 \quad \text{in} \quad C_{\varepsilon,R}, \\ w_\varepsilon = & w \quad \text{on} \quad \Gamma_R, \\ w_\varepsilon = & v_\varepsilon \quad \text{on} \quad \Gamma_\varepsilon, \\ k_2 \Delta v_\varepsilon & = 0 \text{ in } C_{\lambda\varepsilon,\varepsilon}, \end{cases} \tag{2.35}$$

$$\begin{cases} k_2 \partial_{n_2} v_\varepsilon = 0 \text{ on } \Gamma_{\lambda\varepsilon}, \\ k_1 \partial_{n_1} w_\varepsilon + k_2 \partial_{n_2} v_\varepsilon = 0 \text{ on } \Gamma_\varepsilon. \end{cases}$$

The geometry of domains of definition for functions is shown in Fig. 2.3. Now let us adopt the polar coordinate system around origin and assume the Fourier series form for w on Γ_R.

$$w = C_0 + \sum_{k=1}^{\infty} (A_k \cos k\varphi + B_k \sin k\varphi)\,. \tag{2.36}$$

The general form of the solution w_ε is

$$w_\varepsilon = A^w + B^w \log r + \sum_{k=1}^{\infty} \left(w_k^c(r) \cos k\varphi + w_k^s(r) \sin k\varphi \right), \tag{2.37}$$

where

$$w_k^c(r) = A_k^c r^k + B_k^c \frac{1}{r^k}, \quad w_k^s(r) = A_k^s r^k + B_k^s \frac{1}{r^k}.$$

Similarly for v_ε:

$$v_\varepsilon = A^v + B^v \log r + \sum_{k=1}^{\infty} \left(v_k^c(r) \cos k\varphi + v_k^s(r) \sin k\varphi \right), \tag{2.38}$$

where

$$v_k^c(r) = a_k^c r^k + b_k^c \frac{1}{r^k}, \quad v_k^s(r) = a_k^s r^k + b_k^s \frac{1}{r^k}.$$

Additionally, we denote the Fourier expansion of v_ε on Γ_ε by

$$v_\varepsilon = c_0 + \sum_{k=1}^{\infty} (a_k \cos k\varphi + b_k \sin k\varphi) \tag{2.39}$$

From boundary conditions on $\Gamma_{\lambda\varepsilon}$ it follows easily $B^v = 0$, $A^v = c_0$, and then $B^w = 0$, $A^w = A^v = c_0 = C_0$. There remains to find $a_k, b_k, a_k^c, b_k^c, a_k^s, b_k^s, A_k^c, B_k^c, A_k^s, B_k^s$ assuming A_k, B_k as given.

2.4.3 Asymptotic Expansions in Multilayer Subdomain

In order to eliminate the above mentioned coefficients we consider first the terms at $\cos k\varphi$. From boundary and transmission conditions we have for $k = 1, 2, \ldots$

$$\begin{aligned} A_k^c R^k + B_k^c \tfrac{1}{R^k} &= A_k \\ A_k^c \varepsilon^k + B_k^c \tfrac{1}{\varepsilon^k} - a_k &= 0 \\ a_k^c \varepsilon^k + b_k^c \tfrac{1}{\varepsilon^k} - a_k &= 0 \\ a_k^c (\lambda\varepsilon)^{k-1} - b_k^c \tfrac{1}{(\lambda\varepsilon)^{k+1}} &= 0 \\ k_1 A_k^c \varepsilon^{k-1} - k_1 B_k^c \tfrac{1}{\varepsilon^{k+1}} - k_2 a_k^c \varepsilon^{k-1} + k_2 b_k^c \tfrac{1}{\varepsilon^{k+1}} &= 0 \end{aligned} \tag{2.40}$$

This may be rewritten in the matrix form: grouping unknown parameters into a vector $p_k = (A_k^c, B_k^c, a_k^c, b_k^c, a_k)^\top$ we obtain

$$T(k_1, k_2, R, \lambda, \varepsilon) p_k = R^k A_k e_1$$

where

$$T = \begin{bmatrix} R^{2k} & 1 & 0 & 0 & 0 \\ \varepsilon^{2k} & 1 & 0 & 0 & -\varepsilon^k \\ 0 & 0 & (\lambda\varepsilon)^{2k} & 1 & -\varepsilon^k \\ 0 & 0 & (\lambda\varepsilon)^{2k} & -1 & 0 \\ k_1\varepsilon^{2k} & -k_1 & -k_2\varepsilon^{2k} & k_2 & 0 \end{bmatrix} \tag{2.41}$$

where $e_1 = (1, 0, 0, 0, 0)^\top$. It is easy to see that

$$p_k = p_k^0 A_k + \varepsilon^{2k} p_k^1 A_k + o(\varepsilon^{2k}) \tag{2.42}$$

where

$$p_k^0 = \lim_{\varepsilon\to 0+} \lim_{\lambda\to 0+} \frac{p_k(k_1, k_2, R, \lambda, \varepsilon)}{A_k}$$

and $p_k^0 = (1/R^k, 0, 0, 0, 0)^\top$, which corresponds to the ball B_R filled completely with material k_1. Similar reasoning may be conducted for terms containing $\sin k\varphi$. As a result,

$$\mathscr{A}_{\lambda,\varepsilon} = \mathscr{A}_{0,0} + \varepsilon^2 \mathscr{A}^1_{\lambda,\varepsilon}(k_1, k_2, R, \lambda, \varepsilon, A_1, B_1) + o(\varepsilon^2). \tag{2.43}$$

The exact form of $\mathscr{A}^1_{\lambda,\varepsilon}(k_1, k_2, R, \lambda, \varepsilon, A_1, B_1)$ is obtained from inversion of matrix T, but, what is crucial, it is linear in both A_1 and B_1. They in turn are computed as line integrals

$$A_1(w) = \frac{1}{\pi R^2} \int_{\Gamma_R} w x_1 \, ds, \quad B_1(w) = \frac{1}{\pi R^2} \int_{\Gamma_R} w x_2 \, ds.$$

As a result, for computing u_ε we may use the following energy form

$$\mathscr{E}(u_\varepsilon) = \frac{1}{2} \int_\Omega k_1 \nabla u_\varepsilon \cdot \nabla u_\varepsilon \, dx + \varepsilon^2 Q(k_1, k_2, R, \lambda, \varepsilon, A_1, B_1) + o(\varepsilon^2), \tag{2.44}$$

where $A_1 = A_1(u_\varepsilon)$, $B_1 = B_1(u_\varepsilon)$ and Q is a quadratic function of A_1, B_1. This constitutes a regular perturbation of the energy functional which allows computing perturbations of any functional depending on this solution and caused by small inclusion of the described above form.

2.4.4 *Multilayer Subdomains in Linear Elasticity*

Let us consider the plane elasticity problem in the ring $C_{R,\varepsilon}$. We use polar coordinates (r, θ) with e_r pointing outwards and e_θ perpendicularly in the counter-clockwise direction. Then there exists an exact representation of both solutions, using the complex variable series. It has the form [142, 180, 196]

$$\begin{aligned}\sigma_{rr} - i\sigma_{r\theta} &= 2\Re\phi' - e^{2i\theta}(\bar{z}\phi'' + \psi') \\ \sigma_{rr} + i\sigma_{\theta\theta} &= 4\Re\phi' \\ 2\mu(u_r + iu_\theta) &= e^{-i\theta}(\kappa\phi - z\bar{\phi}' - \bar{\psi}).\end{aligned} \tag{2.45}$$

The functions ϕ, ψ are given by complex series

$$\begin{aligned}\phi &= A\log(z) + \sum_{k=-\infty}^{k=+\infty} a_k z^k \\ \psi &= -\kappa\bar{A}\log(z) + \sum_{k=-\infty}^{k=+\infty} b_k z^k.\end{aligned} \tag{2.46}$$

Here μ is the Lame constant, ν is the Poisson ratio, $\kappa = 3 - 4\nu$ in the plain strain case, and $\kappa = (3 - \nu)/(1 + \nu)$ for plane stress.

Similarly as in the simple case described in former sections, the displacement data may be given in the form of Fourier series,

$$2\mu(u_r + iu_\theta) = \sum_{k=-\infty}^{k=+\infty} A_k e^{ik\theta} \tag{2.47}$$

The traction-free condition on some circle means $\sigma_{rr} = \sigma_{r\theta} = 0$. From (2.45), (2.46) we get for displacements the formula

$$\begin{aligned}2\mu(u_r + iu_\theta) = \kappa Ar\log(r)\frac{1}{z} - \bar{A}\frac{1}{r}z + \\ \sum_{p=-\infty}^{p=+\infty}[\kappa r a_{p+1} - (1-p)\bar{a}_{1-p}r^{-2p+1} - \bar{b}_{-(p+1)}r^{-2p-1}]z^p.\end{aligned} \tag{2.48}$$

Similarly we obtain representation of tractions on some circle

$$\begin{aligned}\sigma_{rr} - i\sigma_{r\theta} = 2A\frac{1}{z} + (\kappa+1)\frac{1}{r^2}\bar{A}z + \\ \sum_{p=-\infty}^{p=+\infty}(1-p)\left[(1+p)a_{p+1} + \bar{a}_{1-p}r^{-2p} + \frac{1}{r^2}b_{p-1}\right]z^p.\end{aligned} \tag{2.49}$$

As we see, in principle it is possible to repeat the same procedure again, gluing solutions in two rings together and eliminating the intermediary Dirichlet data on the interface. The only difference lies in considerably more complicated calculations, see e.g. [115]. This could be applied for making double asymptotic expansion, in term of both ε and λ. However, in our case λ does not need to be small in comparison to ε.

Chapter 3
Steklov–Poincaré Operator for Helmholtz Equation

We recall that the topological derivative method is used for the Helmholtz problem [231] already for a long time. The applications of topological derivative method include the inverse problems. The imaging of small acoustic anomalies is investigated in [28, 75, 97, 104, 119, 141]. The experimental verification of the method is presented in [248] for the elastic-wave imaging. In [13] the stability and resolution analysis for the topological-derivative-based imaging functional is given, which confirms that the method is applicable to the inverse scattering [120]. The method is also applied to AFM-based indentation stiffness tomography in [37]. The results on the time domain inverse scattering problems are reported in [54, 79, 84]. The known formulas for the topological derivatives contain the point-wise values of the solutions as well as of the gradients of the solutions for the boundary value problems in the intact or unperturbed geometrical domain. Such values are usually required at the centre of the hole or of the cavity. The numerical evaluation of point-wise values imposes the restrictions on the finite element method used for computations. Therefore, it is useful to obtain an equivalent form of the topological derivative which is written in terms of the solution evaluated in the intact domain and restricted to the boundary of the hole or the cavity. This is possible by an applications of the modified Bessels functions for the Helmholtz equation in two spatial variables.

3.1 Representation of Solutions for Helmholtz Equation

Let us consider the functions satisfying the Helmholtz equation in a certain bigger domain $\Omega \subset \mathbb{R}^2$ and compute the point-wise values of these functions and their derivatives using integrals over circles centered at points in question. The circles constitute boundaries of balls contained in Ω. Our goal is to obtain formulas for point-wise values of solutions and its gradient in the center of these balls, which in turn appear in expansions of Steklov–Poincaré operator. Two distinct cases for the

A. A. Novotny et al., *Applications of the Topological Derivative Method*,
Studies in Systems, Decision and Control 188,
https://doi.org/10.1007/978-3-030-05432-8_3

Helmholtz equation are considered. The first is the case of a coercive bilinear form which we call the coercive Helmholtz operator. In such a case the Lax–Milgram Lemma provides the existence of the unique solution for the associated boundary value problem. In the second case of a non-coercive bilinear form the Fredholm alternative can be used in order to determine the solutions. It is well known that the topological derivative method can be applied for the Helmholtz problems in two spatial dimensions [231]. In general the solutions are complex valued. We restrict ourselves to the real valued solutions for the simplicity. There is no energy functional for the Helmholtz problems except for the case of coercive bilinear form. However, the results for the representation of the Steklov–Poincaré operator can be obtained directly by using the explicit solutions of the Helmholtz equation in the ring.

3.1.1 Case I: Coercive Operator

In this case the Helmholtz equation, which is satisfied in $B_R(\widehat{x}) \subset \Omega$, has the form

$$\begin{cases} -\Delta u_0^\varphi + k^2 u_0^\varphi = 0 \text{ in } B_R(\widehat{x}), \\ u_0^\varphi = \varphi \text{ on } \partial B_R(\widehat{x}). \end{cases} \tag{3.1}$$

The solution, assuming lack of singularity at $\widehat{x}$, may be expressed as

$$u_0^\varphi = \sum_{m=0}^{\infty} \frac{I_m(kr)}{I_m(kR)} \left(A_m \cos m\theta + B_m \sin \theta\right), \tag{3.2}$$

where (r, θ) is a polar coordinates system around the point $\widehat{x}$. Numbers A_m, B_m constitute Fourier coefficients in the expansion of φ on $\partial B_R(\widehat{x})$. Symbols $I_m(x)$ denote modified Bessel functions, which have the following expansions around 0:

$$I_m(x) = \sum_{k=0}^{\infty} \frac{x^{n+2k}}{2^{n+2k} k!(n+k)!} \tag{3.3}$$

All of these functions, with the exception of $I_0(x)$, vanish at $x = 0$ and have no other zeros. In particular

$$I_0(x) = 1 + \frac{1}{2^2} x^2 + \frac{1}{2^4 (2!)^2} x^4 + \ldots$$

$$I_1(x) = \frac{1}{2} x + \frac{1}{2^3 2!} x^3 + \frac{1}{2^5 2! 3!} x^5 + \ldots$$

Let us now assume that $\widehat{x} = 0$, so $x_1 = r\cos\theta$, $x_2 = r\sin\theta$. Then we have

$$r^{m+2k}\cos m\theta = (x_1^2 + x_2^2)^k r^m \cos m\theta = (x_1^2 + x_2^2)^k c_m(x_1, x_2),$$

where $c_m(x_1, x_2)$ is a homogeneous polynomial in x_1, x_2 of the order m. Finally,

$$r^{m+2k}\cos m\theta = \widehat{c}_{m+2k}(x_1, x_2)$$

where $\widehat{c}_{m+2k}(x_1, x_2)$ are also homogeneous polynomials of the order $m + 2k$. For example

$$c_3(x_1, x_2) = x_1^3 - 3x_1x_2^2, \quad \widehat{c}_{3+2k} = (x_1^2 + x_2^2)^k(x_1^3 - 3x_1x_2^2).$$

As a result the solution u_0^φ may be written as

$$u_0^\varphi = \frac{1}{I_0(kR)}A_0 + \frac{k}{2I_1(kR)}(A_1x_1 + B_1x_2) + \mathscr{R}(x_1, x_2). \tag{3.4}$$

Here $\mathscr{R}(x_1, x_2)$ contains second and higher order powers of x_1, x_2. Hence

$$\begin{aligned} u_0^\varphi(0) &= \frac{1}{I_0(kR)}A_0, \\ \frac{\partial u_0^\varphi}{\partial x_1}(0) &= \frac{1}{2I_1(kR)}kA_1, \\ \frac{\partial u_0^\varphi}{\partial x_2}(0) &= \frac{1}{2I_1(kR)}kB_1. \end{aligned}$$

Recalling formulas for Fourier coefficients we get

$$\begin{aligned} A_0 &= \frac{1}{2\pi}\int_0^{2\pi} u_0^\varphi(R,\theta)\,d\theta = \frac{1}{2\pi R}\int_{\partial B_R(0)} u_0^\varphi\,ds, \\ A_1 &= \frac{1}{\pi}\int_0^{2\pi} u_0^\varphi(R,\theta)\cos(\theta)\,d\theta = \frac{1}{\pi R^2}\int_{\partial B_R(0)} u_0^\varphi x_1\,ds, \end{aligned}$$

and similarly

$$B_1 = \frac{1}{\pi R^2}\int_{\partial B_R(0)} u_0^\varphi x_2\,ds.$$

Combining these sets of formulas gives the final result:

$$
\begin{aligned}
u_0^\varphi(0) &= \frac{1}{2\pi R I_0(kR)} \int_{\partial B(0,R)} u_0^\varphi \, ds, \\
\frac{\partial u_0^\varphi}{\partial x_1}(0) &= \frac{k}{2\pi R^2 I_1(kR)} \int_{\partial B(0,R)} u_0^\varphi x_1 \, ds, \\
\frac{\partial u_0^\varphi}{\partial x_2}(0) &= \frac{k}{2\pi R^2 I_1(kR)} \int_{\partial B(0,R)} u_0^\varphi x_2 \, ds.
\end{aligned}
\tag{3.5}
$$

3.1.2 Case II: Non-coercive Operator

This case corresponds to wave-like phenomena and has essentially different character, since operator Δ is negative definite.

$$
\begin{cases}
\Delta u_0^\varphi + k^2 u_0^\varphi = 0 \text{ in } B_R(\widehat{x}), \\
u_0^\varphi = \varphi \text{ on } \partial B_R(\widehat{x}).
\end{cases}
\tag{3.6}
$$

Assuming again the lack of singularity at $\widehat{x}$, we may write down the solution as

$$
u_0^\varphi = \sum_{m=0}^{\infty} \frac{J_m(kr)}{J_m(kR)} (A_m \cos m\theta + B_m \sin\theta) \tag{3.7}
$$

Symbols $J_m(x)$ are Bessel functions and have expansions

$$
I_m(x) = \sum_{k=0}^{\infty} \frac{(-1)^k x^{n+2k}}{2^{n+2k} k!(n+k)!}. \tag{3.8}
$$

There is a fundamental difference with respect to Case I here. Functions $J_n(x)$ have real zeros for $x > 0$ and, as a result, we must take as R such radii that kR is separated from these zeros by a given, fixed margin. It can be done, at least for k of moderate size. All $J_n(x)$, $n \geq 1$, have zero at $x = 0$, with the exception of $J_0(0) = 1$. Other zeros in the vicinity of the origin are only

$$
2.405(j_0), 3.832(J_1), 5.136(J_2), 5.520(J_0), 6.380(J_3), \ldots
$$

Hence if e.g. $kR \in (0, 2) \cup (2.8, 3.4)$, it is separated from zeros of J_n by the margin of 0.4. In practice we should take $kR \in (0, 2)$. Assuming that this is satisfied, and proceeding similarly as in Case I, we have

$$
\begin{aligned}
J_0(x) &= 1 - \frac{1}{2^2} x^2 + \frac{1}{2^4 (2!)^2} x^4 + \ldots \\
J_1(x) &= \frac{1}{2} x - \frac{1}{2^3 2!} x^3 + \frac{1}{2^5 2! 3!} x^5 + \ldots
\end{aligned}
$$

and

$$u_0^\varphi = \frac{1}{J_0(kR)} A_0 + \frac{k}{2J_1(kR)}(A_1 x_1 + B_1 x_2) + \mathscr{R}(x_1, x_2). \tag{3.9}$$

Obviously that gives similar expressions for derivatives

$$u_0^\varphi(0) = \frac{1}{J_0(kR)} A_0,$$
$$\frac{\partial u_0^\varphi}{\partial x_1}(0) = \frac{1}{2J_1(kR)} k A_1,$$
$$\frac{\partial u_0^\varphi}{\partial x_2}(0) = \frac{1}{2J_1(kR)} k B_1.$$

Finally, we obtain analogous formulas

$$\begin{aligned} u_0^\varphi(0) &= \frac{1}{2\pi R J_0(kR)} \int_{\partial B(0,R)} u_0^\varphi \, ds, \\ \frac{\partial u_0^\varphi}{\partial x_1}(0) &= \frac{k}{2\pi R^2 J_1(kR)} \int_{\partial B(0,R)} u_0^\varphi x_1 \, ds, \\ \frac{\partial u_0^\varphi}{\partial x_2}(0) &= \frac{k}{2\pi R^2 J_1(kR)} \int_{\partial B(0,R)} u_0^\varphi x_2 \, ds. \end{aligned} \tag{3.10}$$

3.2 Numerical Testing of Approximate Formulas for Steklov–Poincaré Operators

In order to check if there is no errors in coefficients appearing in the above expressions, we shall use the exemplary solutions. The point of origin will be $\widehat{x} = (3, 2)$, $R = 0.9$ and $k = 1.1$ in both cases. These non-integer values are designed to avoid spurious coincidences. The solutions used for testing will be

$$u_1 = \frac{I_3(kr)}{I_3(5k)}(2\cos 3\theta + \sin 3\theta)$$

in Case I and

$$u_2 = J_3(kr)(2\cos 3\theta + \sin 3\theta)$$

in Case II. Here (r, θ) are polar coordinates centered at $(3, 2)$. Both functions are shown in Fig. 3.1. We compute $u_1(0)$ and $u_2(0)$ both analytically and numerically using our formulas. Analytical computations are easy using recurrence relations for Bessel functions. The results are shown in Table 3.1.

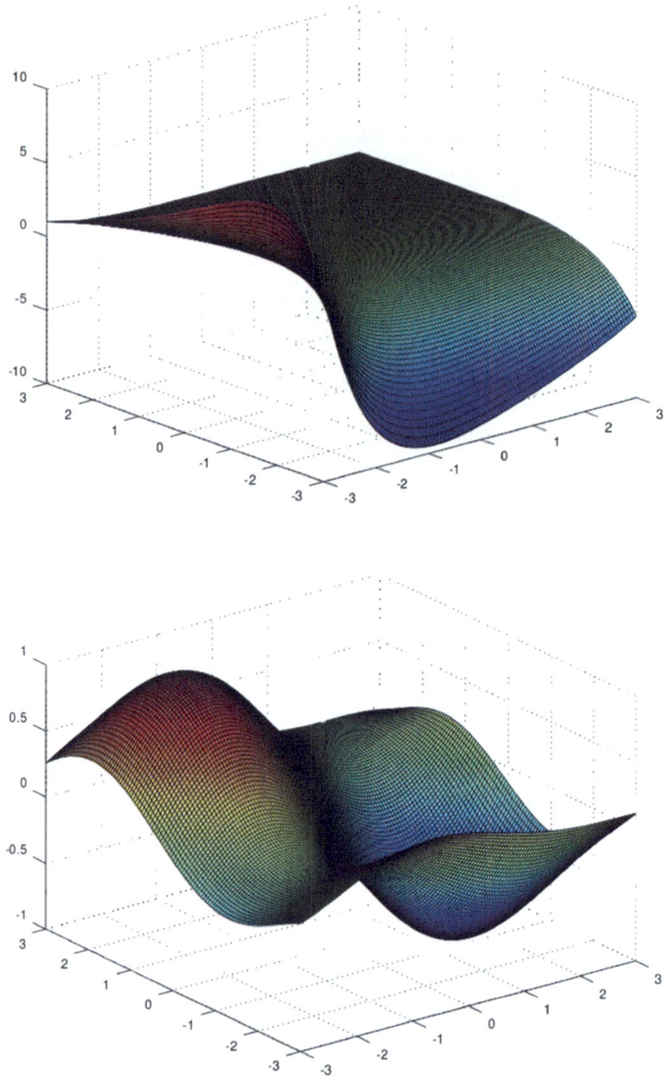

Fig. 3.1 Graphs of solutions u_1 (top) and u_2 (bottom)

Table 3.1 Values of solutions $u_1(0)$ and $u_2(0)$

Values	$u_1(0)$	$u_2(0)$
Analytical	0.29617797	0.45264481
Numerical	0.29617801	0.45264470

3.3 Solutions in the Ring for Helmholtz

Let us now consider two problems concerning Case II defined in the ring $C_{\varepsilon,R} = \{x : \varepsilon < \|x\| < R\}$ with boundaries $\Gamma_R = \{x : \|x\| = R\}$ and $\Gamma_\varepsilon = \{x : \|x\| = \varepsilon\}$. The Neumann problem reads:

$$\begin{cases} \Delta u^{\varphi}_{N,\varepsilon} + k^2 u^{\varphi}_0 = 0 \text{ in } C_{\varepsilon,R}, \\ u^{\varphi}_{N,\varepsilon} = \varphi \text{ on } \Gamma_R, \\ \partial_n u^{\varphi}_{N,\varepsilon} = 0 \text{ on } \Gamma_\varepsilon. \end{cases} \tag{3.11}$$

On the other hand, the Dirichlet problem is stated as:

$$\begin{cases} \Delta u^{\varphi}_{D,\varepsilon} + k^2 u^{\varphi}_0 = 0 \text{ in } C_{\varepsilon,R}, \\ u^{\varphi}_{D,\varepsilon} = \varphi \text{ on } \Gamma_R, \\ u^{\varphi}_{D,\varepsilon} = 0 \text{ on } \Gamma_\varepsilon. \end{cases} \tag{3.12}$$

They have solutions

$$u^{\varphi}_{N,\varepsilon} = \sum_{m=0}^{\infty} (A_m \cos m\theta + B_m \sin\theta)\left(C^N_m J_m(kr) + D^N_m Y_m(kr)\right) \tag{3.13}$$

for Neumann case and

$$u^{\varphi}_{D,\varepsilon} = \sum_{m=0}^{\infty} (A_m \cos m\theta + B_m \sin\theta)\left(C^D_m J_m(kr) + D^D_m Y_m(kr)\right) \tag{3.14}$$

for the Dirichlet case. The coefficients $C^N_m, D^N_m, C^D_m, D^D_m$ may be easily computed from conditions:

$$\begin{aligned} C^N_m J_m(kR) + D^N_m Y_m(kR) &= 1, \\ C^N_m \frac{d}{dr} J_m(k\varepsilon) + D^N_m \frac{d}{dr} Y_m(k\varepsilon) &= 0, \end{aligned}$$

for the Neumann problem and

$$\begin{aligned} C^D_m J_m(kR) + D^D_m Y_m(kR) &= 1, \\ C^D_m J_m(k\varepsilon) + D^D_m Y_m(k\varepsilon) &= 0, \end{aligned}$$

for the Dirichlet problem.

Remark 3.1 The above approach would work only for long waves, i.e. small k. It is the consequence of the requirement, that $kR < 2 \Longrightarrow R < 2/k$, $k\varepsilon < 0.8 \Longrightarrow \varepsilon < 0.8/k$. The numbers are related to roots of Bessel functions: first root of $J_n(x), n \geq 1$, is ≈2.41, and first root of $Y_n(x)$, $n \geq 0$ is ≈0.89. So the condition for R limits the practicability of the approach.

Assuming that the above bounds for k, R hold, and using the formulas

$$J'_m = 0.5(J_{m-1} - J_{m+1}), \quad J_{-m} = -J_m$$

$$Y_m = 0.5(Y_{m-1} - Y_{m+1}), \quad Y_{-m} = -Y_m$$

we can easily solve these systems and compare the obtained coefficients to the case of full ball. These differences are:

$$\begin{aligned} CN0 &= C_0^N - 1/J_0(kR) \\ DN0 &= D_0^N \\ CN1 &= C_1^N - 1/J_1(kR) \\ DN1 &= D_1^N \end{aligned}$$

for the Neumann case and

$$\begin{aligned} CD0 &= C_0^D - 1/J_0(kR) \\ DD0 &= D_0^D \end{aligned}$$

for the Dirichlet problem. In the latter case, as we shall see, the principal part of the expansion in terms of functions of ε appears only in first differences.

It may be checked computationally, assuming $k = 0.9$ and $R = 1.1$, and using series of ε of the form

$$\varepsilon_i = 0.2\, R\, 2^{-i}, \quad i = 1, \ldots 20$$

that

$$\begin{aligned} |CN0| &= C_1(k, R)\varepsilon^2 + O(\varepsilon^4) \\ |DN0| &= C_2(k, R)\varepsilon^2 + O(\varepsilon^4) \\ |CN1| &= C_3(k, R)\varepsilon^2 + O(\varepsilon^4) \\ |DN1| &= C_4(k, R)\varepsilon^2 + O(\varepsilon^4) \\ |CD0| &= C_5(k, R)|\log\varepsilon|^{-1} + O(\varepsilon^3) \\ |DD0| &= C_6(k, R)|\log\varepsilon|^{-1} + O(\varepsilon^3) \end{aligned}$$

This is illustrated in Fig. 3.2, where we see graphs for the Neumann case, and in Fig. 3.3, showing Dirichlet case.

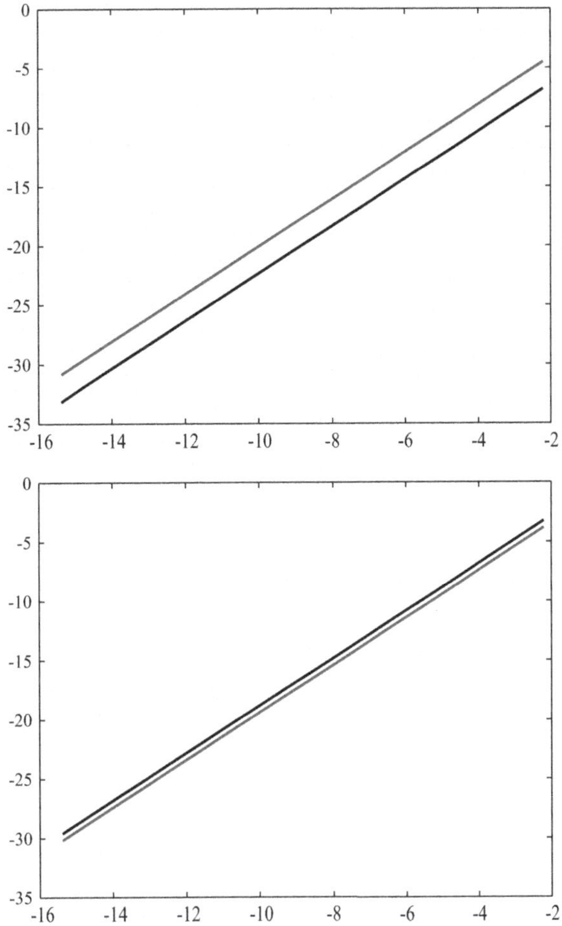

Fig. 3.2 On the top graphs of $\log(|C_0^N - 1/J_0(kR)|)$ and $\log(|D_0^N|)$, on the bottom $\log(|C_1^N - 1/J_1(kR)|)$ and $\log(|D_1^N|)$, always against $\log(\varepsilon_i)$

3.4 Precision of Formulas for Helmholtz in Both Cases

The formulas for Case I are of immediate use, since topological derivative for such problems are expressed in terms of point-wise values of solutions and their gradients. Therefore, the replacement of these values by integrals over circles allows extension of the approach based on topological derivative also to problems with unilateral constraints for solutions.

The Case II requires satisfying the conditions for R, k mentioned above. This does not change the theoretical correctness of expansions of solutions to the equations in the ring of the form

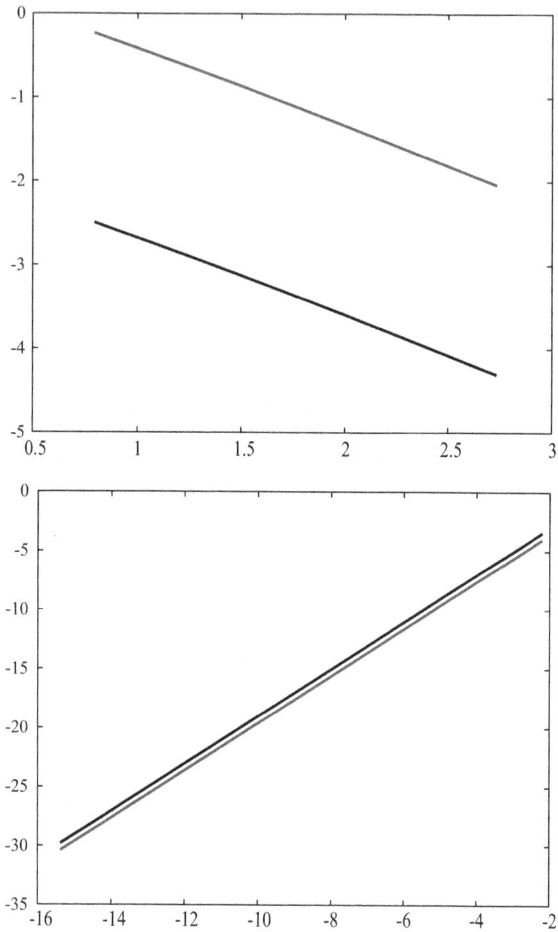

Fig. 3.3 On the top graphs of $\log(|C_0^D - 1/J_0(kR)|)$ and $\log(|D_0^D|)$ against $\log(|\log(\varepsilon_i)|)$, on the bottom $\log(|C_1^D - 1/J_1(kR)|)$ and $\log(|D_1^D|)$, against $\log(\varepsilon_i)$

$$u^{\varphi}_{D,\varepsilon} = u^{\varphi}_0 + c_1(k, R)u^{\varphi}_0(0)\frac{1}{|\log \varepsilon|} + O(\varepsilon^3)$$

$$u^{\varphi}_{N,\varepsilon} = u^{\varphi}_0 + c_2(k, R)u^{\varphi}_0(0)\varepsilon^2 + c_3(k, R)\frac{\partial u^{\varphi}_0}{\partial x_1}(0)\varepsilon^2 + c_4(k, R)\frac{\partial u^{\varphi}_0}{\partial x_2}(0)\varepsilon^2 + O(\varepsilon^4)$$

Since values of k and R are fixed, the constants $C_i(k, R)$, as well as $c_i(k, R)$ may be easily obtained numerically; they can also be computed exactly from explicit forms of series for Bessel functions and (3.10). The practical difficulty arises due to requirement that $R < 2/k$, which imposes necessity of using very fine discretization for big k. But this necessity in consistent with physical interpretation (short waves). With these caveats, topological derivatives and Steklov–Poincaré expansions work in both cases.

Chapter 4
Topological Derivatives for Optimal Control Problems

In this chapter, the topological derivative is obtained for the optimal value of the cost functional for a class of optimal control problems. We obtain the closed form of the topological derivative. Such an approach would allow to consider simultaneous structure design and control modifications. In order to introduce the topological derivative of the shape functional $\Omega \to J(\Omega)$ we usually consider the mapping $\Omega \to y$ for the boundary value problem which gives the function y. In the case of optimal control problems the couple (u, y) of the control u and the state y is given by the optimality system. If the state equation is defined in Ω then the optimal couple is a function of the domain, say $\Omega \to (u^\star, y^\star)$. Therefore, we consider the optimal control problem with the state equation defined in the domain Ω. Hence, the optimal control $u^\star$ is given by the minimization of the cost functional $u \to I(u)$ for fixed Ω, and the optimal state $y^\star$ is obtained by the solution of the state equation in Ω. The optimal value of the cost $I(u^\star)$ defines a domain functional $\Omega \to J(\Omega) := I(u^\star)$ depending on the domain of integration of the elliptic state equation.

4.1 Example in One Spatial Dimension

It is useful to introduce the topological derivative for a control problem in one spatial dimension. First, the state equation is introduced. The state is denoted by $x \to y = y(x), x \in \Omega :=]-1, 1[$ and it is given by the unique solution of the elliptic equation

$$-y''(x) + \chi_\rho(x)y(x) = u(x) \quad \text{in} \quad \Omega; \qquad y(-1) = u_1,\ y(1) = u_2$$

There are triplet controls $u = (u_1, u_2, u) \in \mathbb{R}^2 \times L^2(-1, 1)$. The state equation depends on the first small parameter $\rho \to 0$ since $x \to \chi_\rho(x)$ is a characteristic function of the interval $]x_0 - \rho/2, x_0 + \rho/2[\subset \Omega$, where $x_0 \in \Omega$ is given point. The cost functional is of tracking type

A. A. Novotny et al., *Applications of the Topological Derivative Method*,
Studies in Systems, Decision and Control 188,
https://doi.org/10.1007/978-3-030-05432-8_4

$$I(u_1, u_2, u) = \frac{1}{2}\int_{-1}^{1}(y(x) - z_d(x))^2 + \frac{1}{2}\|(u_1, u_2, u)\|^2,$$

where $z_d \in L^2(-1, 1)$ is a given function and $(u_1, u_2, u) \to \|(u_1, u_2, u)\|$ is a norm on the space $\mathbb{R}^2 \times L^2(-1, 1)$. The optimal control which minimizes the cost is denoted by $(u_1^\star, u_2^\star, u^\star)$. The control is given by the optimality system. Hence, we define the domain functional

$$J(\Omega) := I(u^\star) = \frac{1}{2}\int_{-1}^{1}(y^\star(x) - z_d(x))^2 + \frac{1}{2}\|(u_1^\star, u_2^\star, u^\star)\|^2, \tag{4.1}$$

Now we introduce the small parameter $\varepsilon \to 0$ which governs the singular perturbation of the geometrical domain. In order to introduce the topological perturbation Ω_ε depending on the small parameter $\varepsilon \to 0$ of the domain $\Omega :=]-1, 1[$ we fix a point $\widehat{x} \in \Omega$ and define $\Omega_\varepsilon = \Omega \setminus [\widehat{x} - \varepsilon/2, \widehat{x} + \varepsilon/2]$. The state equation reads

$$-y_\varepsilon''(x) + \chi_\rho(x) y_\varepsilon(x) = u(x) \quad \text{in} \quad \Omega_\varepsilon.$$

We add the Dirichlet homogeneous boundary conditions at the points $\widehat{x} - \varepsilon/2, \widehat{x} + \varepsilon/2 \in \Omega$ for the state equation,

$$y_\varepsilon(-1) = u_1, \ y_\varepsilon(1) = u_2, \ y_\varepsilon(\widehat{x} - \varepsilon/2) = y_\varepsilon(\widehat{x} - \varepsilon/2) = 0.$$

We are interested in the expansion

$$J(\Omega_\varepsilon) = J(\Omega) + f(\varepsilon)\mathscr{T}(\widehat{x}) + o(f(\varepsilon)). \tag{4.2}$$

The state equation depends on the small parameter $\varepsilon \to 0$, the first possible question is the asymptotic analysis of the solutions and of the control problem with respect to this parameter. Thus, we consider the expansion of the optimal value of the cost with respect to the small parameter, which we leave to the interested reader as an exercise.

4.2 Control Problem

Let us consider the domain Ω in $\mathbb{R}^2$ with piecewise smooth boundary, its subset D (also with the piecewise smooth boundary) and the control problem governed by the elliptic state equation,

$$\begin{cases} -\Delta y = \chi_D\, u \ \text{in} \ \Omega, \\ \quad\ \ y = 0 \qquad \text{on} \ \Gamma = \partial\Omega. \end{cases}$$

For given $u \in L^2(D)$, $y = y(u)$ represents e.g. the deflection of an elastic membrane, loaded by the vertical force u concentrated on D. The characteristic function of D is denoted χ_D, and Ω is the reference domain for the membrane. For such a system we define the cost functional

$$I(u) = \mathscr{I}(y(u), u) = \int_D F(y, u) d\Omega = \frac{1}{2} \int_D [(y - y_0)^2 + \alpha u^2] d\Omega,$$

which is minimized over the space of controls $u \in L^2(D), \alpha > 0$, where y_0 is a given function. The minimization of $I(u)$ with respect to u means approximation of a given function y_0 in the region D by the deflection (shape) of an elastic membrane, using the smallest possible load u applied in D. The extremal value of the cost functional for this control problem defines the shape functional, depending on the geometrical domain Ω,

$$\mathscr{J}(\Omega) = \min_{u \in L^2(D)} \mathscr{I}(y(u), u).$$

Variation of the state $y'(v)$, corresponding to the variation v of the control

$$y(u + sv) = y(u) + sy'(v) ,$$

satisfies the equation

$$\begin{cases} -\Delta y' = \chi_D \ v \ \text{in} \ \Omega, \\ \quad y' = 0 \qquad \text{on} \ \Gamma, \end{cases}$$

and the variation $\delta I(u; v)$ of the cost I is given by

$$\delta I(u; v) = \int_D [F_y(y, u) \ y'(v) + F_u(y, u) \ v] \, d\Omega.$$

Introducing the adjoint equation

$$\begin{cases} -\Delta p = \chi_D \ F_y(y, u) \ \text{in} \ \Omega, \\ \quad p = 0 \qquad\qquad\quad \text{on} \ \Gamma, \end{cases}$$

allows us to express the first term in the cost variation as follows

$$\begin{aligned} \int_D F_y \ y' \, d\Omega &= - \int_\Omega \Delta p \ y' \, d\Omega = \int_\Omega \nabla p \cdot \nabla y' \, d\Omega \\ &= - \int_\Omega p \ \Delta y' \, d\Omega = \int_D pv \, d\Omega. \end{aligned}$$

Hence

$$\delta I(u; v) = \int_D [p + F_u(y, u)] \, v \, d\Omega$$

and the stationarity condition

$$\delta I(u; v) = 0\,, \quad \forall v \in L^2(D)$$

takes on the form

$$F_u(y(u; x), u(x)) = -p(x) \quad \text{a.e. in} \quad D\,.$$

For the specific choice of the cost functional, this results in the equality $u = -\frac{1}{\alpha}p$ and gives the extremal value of the cost functional for the control problem in the following form

$$\mathscr{J}(\Omega) = \frac{1}{2}\int_D \left[(y - y_0)^2 + \frac{1}{\alpha}p^2\right] d\Omega,$$

where y, p are given as a solution of the coupled system of equations:

$$\begin{cases} -\Delta y = -\chi_D\,\frac{1}{\alpha}p & \text{in } \Omega, \\ y = 0 & \text{on } \Gamma, \\ p = 0 & \text{on } \Gamma. \end{cases}$$

4.3 Topological Derivative

The variation of the geometrical domain Ω resulting in the change of the topological characteristic consists in removing a small ball centered at the point $x_0 \in \operatorname{int}(\Omega \setminus D)$, such, that $B_\varepsilon(x_0) \subset \Omega \setminus D$ for sufficiently small $0 < \varepsilon < \varepsilon_0$. Denoting $\Omega_\varepsilon = \Omega \setminus B_\varepsilon(x_0)$, we define the optimal value of the cost functional for the control problem defined in the domain of integration Ω_ε as a function depending on small parameter $\varepsilon > 0$ and the point $x_0 \in \Omega$,

$$J(\varepsilon) = J(\varepsilon; x_0) = \frac{1}{2}\int_D \left[(y_\varepsilon - y_0)^2 + \frac{1}{\alpha}p_\varepsilon^2\right] d\Omega.$$

The state and adjoint variables y_ε, p_ε are given by the unique solutions to the following optimality system,

$$\begin{cases} -\Delta y_\varepsilon = -\chi_D\,\frac{1}{\alpha}p_\varepsilon & \text{in } \Omega_\varepsilon\,, \\ -\Delta p_\varepsilon = \chi_D\,(y_\varepsilon - y_0) & \text{in } \Omega_\varepsilon\,, \\ y_\varepsilon = 0 & \text{on } \Gamma, \\ p_\varepsilon = 0 & \text{on } \Gamma, \\ \partial_n y_\varepsilon = 0 & \text{on } \Gamma_\varepsilon, \\ \partial_n p_\varepsilon = 0 & \text{on } \Gamma_\varepsilon, \end{cases}$$

where we denote $\Gamma_\varepsilon = \partial B_\varepsilon(x_0)$. Observe, that we have imposed the free edge condition on the boundaries of holes.

Our objective is to analyse the behaviour of $J(\varepsilon) = \mathscr{J}(\Omega_\varepsilon)$ as $\varepsilon \to 0+$. To this end we evaluate the limits of derivatives $J'(\varepsilon)$, $J''(\varepsilon)$ for $\varepsilon \downarrow 0$. Using the formulae given in [238, 245] the following form of the shape derivative of $\mathscr{J}(\Omega_\varepsilon)$ is obtained,

$$J'(\varepsilon) = \lim_{s\to 0} \frac{1}{s}(J(\varepsilon + s) - J(\varepsilon)) = \int_D \left[(y_\varepsilon - y_0)y'_\varepsilon + \frac{1}{\alpha} p_\varepsilon p'_\varepsilon \right] d\Omega,$$

where y'_ε, p'_ε are strong shape derivatives [245] of solutions y_ε, p_ε to the state equation and the adjoint state equation, respectively. The shape derivatives y'_ε, p'_ε satisfy the following equations in the weak forms:

$$\int_{\Omega_\varepsilon} \left[\nabla y'_\varepsilon \cdot \nabla \phi_1 + \frac{1}{\alpha} \chi_D \; p'_\varepsilon \phi_1 \right] d\Omega = \int_{\Gamma_\varepsilon} \frac{\partial y_\varepsilon}{\partial \tau} \frac{\partial \phi_1}{\partial \tau} d\Gamma$$

$$\int_{\Omega_\varepsilon} [\nabla p'_\varepsilon \cdot \nabla \phi_2 - \chi_D \; y'_\varepsilon \phi_2] d\Omega = \int_{\Gamma_\varepsilon} \frac{\partial p_\varepsilon}{\partial \tau} \frac{\partial \phi_2}{\partial \tau} d\Gamma$$

for all test functions $\phi_1, \phi_2 \in H^1_0(\Omega) \cap H^2(\Omega)$, where $\partial/\partial\tau$ denotes the derivative in the tangential direction to the boundary of the hole. In order to simplify the form of the derivative $J'(\varepsilon)$, we can introduce the second level adjoint variables ξ_ε, η_ε, defined by the following system of equations:

$$\begin{cases} -\Delta \xi_\varepsilon - \chi_D \; \eta_\varepsilon = \chi_D \; (y_\varepsilon - y_0) & \text{in } \Omega_\varepsilon \, , \\ -\Delta \eta_\varepsilon + \frac{1}{\alpha} \chi_D \; \xi_\varepsilon = \frac{1}{\alpha} \chi_D \; p_\varepsilon & \text{in } \Omega_\varepsilon \, , \\ \xi_\varepsilon = 0 & \text{on } \Gamma, \\ \eta_\varepsilon = 0 & \text{on } \Gamma, \\ \partial_n \xi_\varepsilon = 0 & \text{on } \Gamma_\varepsilon \, , \\ \partial_n \eta_\varepsilon = 0 & \text{on } \Gamma_\varepsilon \, , \end{cases}$$

or, in the weak form, with the test functions $\psi_1, \psi_2 \in H^1_0(\Omega)$:

$$\int_{\Omega_\varepsilon} [\nabla \xi_\varepsilon \cdot \nabla \psi_1 - \chi_D \; \eta_\varepsilon \psi_1] d\Omega = \int_D (y_\varepsilon - y_0)\psi_1 \, d\Omega,$$

$$\int_{\Omega_\varepsilon} \left[\nabla \eta_\varepsilon \cdot \nabla \psi_2 + \frac{1}{\alpha} \chi_D \; \xi_\varepsilon \psi_2 \right] d\Omega = \int_D \frac{1}{\alpha} p_\varepsilon \psi_2 \, d\Omega.$$

After substituting $\phi_1 := \xi_\varepsilon$, $\phi_2 := \eta_\varepsilon$, $\psi_1 := y'_\varepsilon$, $\psi_2 := p'_\varepsilon$, these integral identities take on the forms

$$\int_{\Omega_\varepsilon} \left[\nabla y'_\varepsilon \cdot \nabla \xi_\varepsilon + \frac{1}{\alpha} \chi_D \; p'_\varepsilon \xi_\varepsilon \right] d\Omega = \int_{\Gamma_\varepsilon} \frac{\partial y_\varepsilon}{\partial \tau} \frac{\partial \xi_\varepsilon}{\partial \tau} d\Gamma,$$

$$\int_{\Omega_\varepsilon} [\nabla p'_\varepsilon \cdot \nabla \eta_\varepsilon - \chi_D \; y'_\varepsilon \eta_\varepsilon] d\Omega = \int_{\Gamma_\varepsilon} \frac{\partial p_\varepsilon}{\partial \tau} \frac{\partial \eta_\varepsilon}{\partial \tau} d\Gamma,$$

$$\int_{\Omega_\varepsilon} [\nabla \xi_\varepsilon \cdot \nabla y'_\varepsilon - \chi_D \, \eta_\varepsilon y'_\varepsilon] d\Omega = \int_D (y_\varepsilon - y_0) y'_\varepsilon \, d\Omega,$$

$$\int_{\Omega_\varepsilon} \left[\nabla \eta_\varepsilon \cdot \nabla p'_\varepsilon + \frac{1}{\alpha} \chi_D \, \xi_\varepsilon p'_\varepsilon \right] d\Omega = \int_D \frac{1}{\alpha} p_\varepsilon p'_\varepsilon \, d\Omega,$$

and as a result we get the following expression for the first order shape derivative of the functional $\varepsilon \to \mathcal{J}(\Omega_\varepsilon) = J(\varepsilon)$:

$$J'(\varepsilon) = \int_{\Gamma_\varepsilon} \left(\frac{\partial y_\varepsilon}{\partial \tau} \frac{\partial \xi_\varepsilon}{\partial \tau} + \frac{\partial p_\varepsilon}{\partial \tau} \frac{\partial \eta_\varepsilon}{\partial \tau} \right) d\Gamma = \int_{\Gamma_\varepsilon} G \, d\Gamma.$$

In order to obtain the limit of $J'(\varepsilon)$ for $\varepsilon \downarrow 0$, we use the asymptotic expansions for the solutions of elliptic equations in the neighbourhood of the small circular hole with respect to the radius of the hole. Observe, that in the case of our control problem, the functions y_ε, p_ε are not solutions to single equations, but to a coupled system of equations. However, y_ε, p_ε are harmonic outside of D, and it can be shown, that the same type of expansions as obtained for the single equation can be derived for y_ε, p_ε. This can be expressed in the following way. Let

$$\nabla y(x_0) = [a, b]^T \,, \quad x_0 \in \Omega \setminus D \,.$$

The solution y_ε as a function of polar coordinates r, θ in an neighbourhood of the ball $B_\varepsilon(x_0)$, can be expressed for $r \geq \varepsilon$ as follows:

$$y_\varepsilon = y + a \frac{\varepsilon^2}{r} \cos\theta + b \frac{\varepsilon^2}{r} \sin\theta + \mathcal{R}$$

where

$$\mathcal{R} = \varepsilon^2 \left[O\left(\frac{\varepsilon}{r}\right) + l(\varepsilon, r) \right],$$

and $l(\varepsilon, r)$ may contain finite powers of $\ln \varepsilon$, $\ln r$. Hence $\mathcal{R} = O(\varepsilon^{2-\delta})$ for any $\delta > 0$. Therefore, in the ring $\varepsilon \leq r \leq 2\varepsilon$, taking into account the regularity of $y = y_0$ in the neighbourhood of x_0 and using the Taylor expansion for y, we have the following expansion for y_ε,

$$y_\varepsilon = y(x_0) + a \left(\frac{\varepsilon^2}{r} + r \right) \cos\theta + b \left(\frac{\varepsilon^2}{r} + r \right) \sin\theta + O(\varepsilon^{2-\delta}),$$

where $y(x_0)$ denotes the value at x_0 of the solution to the elliptic equation in the domain Ω, i.e., in the full domain without hole.

The above formulae are given in the polar coordinate system with the center at x_0, which coincides with the center of the ball. In particular, from the expansion it follows that

$$\frac{\partial y_\varepsilon}{\partial \tau} = \frac{1}{\varepsilon}\frac{\partial y_\varepsilon}{\partial \theta}|_{r=\varepsilon} = 2(-a\sin\theta + b\cos\theta) + O(\varepsilon^{1-\delta}).$$

Now, using these expansions for $y_\varepsilon, p_\varepsilon, \xi_\varepsilon, \eta_\varepsilon$, it can be shown, that

$$\lim_{\varepsilon\to 0+} J'(\varepsilon) = 0,$$

and therefore we evaluate the second derivative $J''(\varepsilon)$. It should be noted, that the existence of asymptotic expansions for functions $\xi_\varepsilon, \eta_\varepsilon$ requires a separate proof, since in the system of equations for $\xi_\varepsilon, \eta_\varepsilon$ not only the geometrical domains of integration, but also right-hand sides depend on the parameter ε. Using the appropriate formulae given e.g. in [238, 245] leads to the following form of the second order derivative

$$J''(\varepsilon) = \int_{\Gamma_\varepsilon} \left(\frac{d}{d\varepsilon}G - \frac{\partial}{\partial n}G\right) d\Gamma + \frac{1}{\varepsilon}\int_{\Gamma_\varepsilon} G\, d\Gamma.$$

Again, it follows, that the first integral vanishes as $\varepsilon \to 0+$ and we obtain

$$J''(0) = \lim_{\varepsilon\to 0+} J''(\varepsilon) = \lim_{\varepsilon\to 0+} \frac{1}{\varepsilon}\int_{\Gamma_\varepsilon} \left(\frac{\partial y_\varepsilon}{\partial \tau}\frac{\partial \xi_\varepsilon}{\partial \tau} + \frac{\partial p_\varepsilon}{\partial \tau}\frac{\partial \eta_\varepsilon}{\partial \tau}\right) d\Gamma.$$

Using once more the asymptotic expansions and the explicit form of the above integral allows us to perform the passage to the limit which results in the final formula for the second order derivative at $\varepsilon = 0^+$.

Theorem 4.1 *The topological derivative*

$$\mathcal{T}(x_0) = \lim_{\varepsilon\to 0^+} \frac{\mathcal{J}(\Omega \setminus B_\varepsilon(x_0)) - \mathcal{J}(\Omega)}{|B_\varepsilon(x_0)|}$$

is given by the following formula

$$\mathcal{T}(x_0) = \frac{J''(0; x_0)}{2\pi} = 2(\nabla y(x_0)\cdot\nabla\xi(x_0) + \nabla p(x_0)\cdot\nabla\eta(x_0)),$$

where y, p, ξ, η denotes the solutions to the equations defined in Ω.

The formula gives the map of the second derivative of the functional $J(\varepsilon)$ at $\varepsilon = 0^+$ as a function of the point $x_0 \in \Omega$. The negative value of $J''(0; x_0)$ indicates, that removing from Ω a sufficiently small ball around x_0 would result in decreasing of the optimal value $\mathcal{J}(\Omega)$ of the cost functional for control problem, giving rise to the new, improved design, but with different topological characteristics.

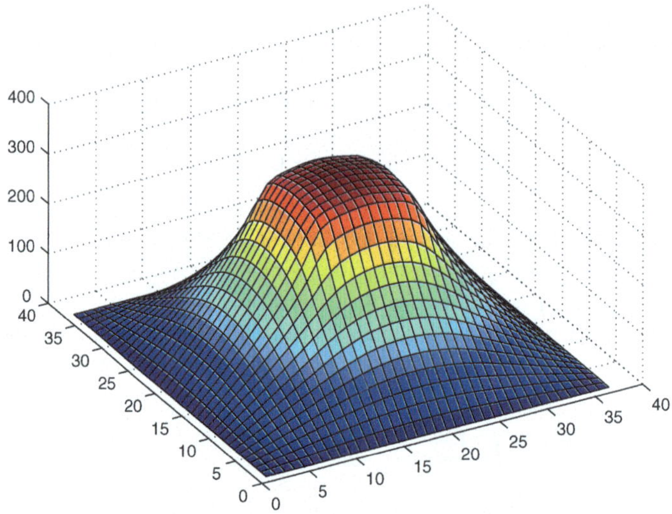

Fig. 4.1 The solution y

4.4 Numerical Example

We consider the domain $\Omega = [-3, 3] \times [-3, 3]$ and its subset $D = [-1, 1] \times [-1, 1]$. The boundary conditions are the same as in the previous sections, and the reference solution is

$$y_0 = 9 - r^2.$$

The control penalty parameter takes value $\alpha = 0.2$. In Fig. 4.1 we see the solution, which should on the central square approximate y_0. Indeed, the graph of the difference $y - y_0$ in Fig. 4.2 confirms this. The control value u is presented in Fig. 4.3. The contour map of $J''(0; x)$ is shown in Fig. 4.4. The darker the shade, the smaller the value of the function. Noting the dark patches indicate where the material should be weakened in order to decrease the extreme value of the goal functional (better accuracy at smaller cost). Observe, that they coincide with the regions of high control value (Fig. 4.3), as should be expected.

4.5 Final Remarks

The methodology described in this chapter works well in examples, where we can get the asymptotic expansions of the solution to the state equation. However, this is not always easy. For example, if $D = \Omega$ in our case, the functions y_ε, p_ε cease to be harmonic around holes, since they satisfy the system of equations

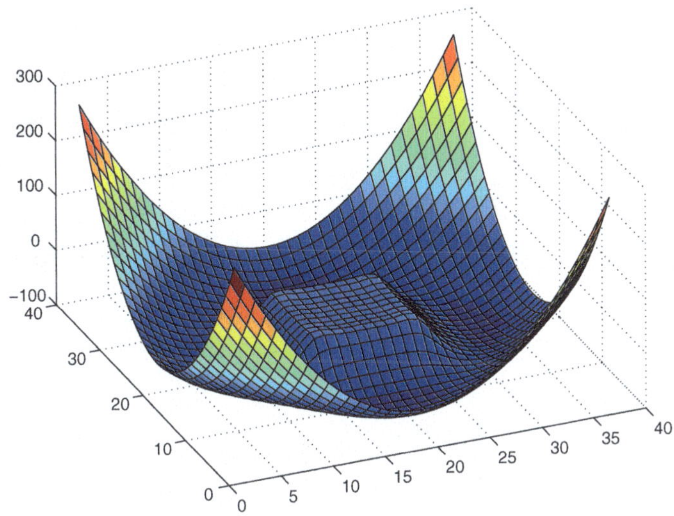

Fig. 4.2 The difference between solution y and y_0

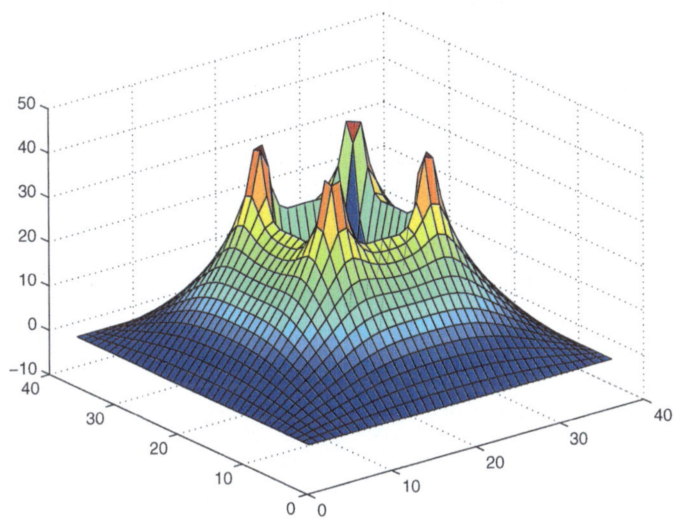

Fig. 4.3 The value of the control u

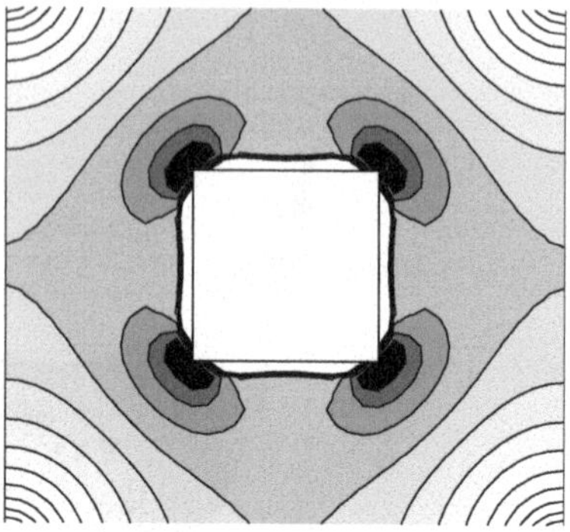

Fig. 4.4 The contour map of $J''(0;x)$

$$\begin{cases} -\Delta y = -\frac{1}{\alpha}p & \text{in } \Omega, \\ -\Delta p = (y - y_0) & \text{in } \Omega, \\ y = 0 & \text{on } \Gamma, \\ p = 0 & \text{on } \Gamma, \end{cases}$$

which may be transformed for the single variable to

$$\Delta^2 p + k^2 p = 0.$$

The asymptotic expansions are in this case much more complicated, involving Jacobi functions. We have presented the simpler case in order not to obscure the idea of the application of topological derivative in control problems.

Chapter 5
Optimality Conditions with Topological Derivatives

In classical theory of shape optimization the first order necessary optimality conditions account for boundary variations of an optimal domain. On the other hand the relaxed formulation based on homogenization technique is used [5, 46, 173] in the topology optimization of energy functionals, the so called compliance in structural optimization. For such a formulation the coefficients of an elliptic operator are selected in an optimal way and the resulting optimal design takes the form of a composite microstructure rather than any geometrical domain. On the other hand the so-called *bubble method* is used for the topology optimization in structural mechanics [92, 233], which leads to numerical methods. We refer also to [105, 139, 172] for the related results. Further applications in mechanics can be found in [139, 169–171, 250]. It seems that in the literature on the subject there is a lack of general method or technique that can be applied in the process of optimization of an arbitrary shape functional for simultaneous boundary and topology variations. Such an approach would be very useful for numerical solution of e.g. optimum design problems in structural mechanics. In the paper [238] the so called topological derivative (TD) of an arbitrary shape functional is introduced. Such a derivative is evaluated by an application of the asymptotic analysis with respect to geometrical singularities of domains [134, 185, 186] for a class of elliptic equations including 2D elasticity system [238] and 3D elasticity system [240]. TD determines weather a change of topology by creation of a small hole, or in similar setting of a small inclusion, at a given point $x \in \Omega$ would result in improving the value of the given shape functional. In general, the form of topological derivatives is established [238, 240] by using the asymptotic expansions of solutions to elliptic systems obtained by the method of matched (or compound) asymptotics. In the case of cavities in the form of two dimensional circles or three dimensional balls (in the case of Laplace equations in the form of balls in $\mathbb{R}^d$ for an arbitrary space dimension $d \geq 2$) the constructive results are obtained [238, 240] by using the shape calculus combined with the asymptotic expansions of solutions.

A. A. Novotny et al., *Applications of the Topological Derivative Method*,
Studies in Systems, Decision and Control 188,
https://doi.org/10.1007/978-3-030-05432-8_5

5.1 Preliminaries

In the present chapter the approach of [238] is extended to the case of a finite number of circular holes treated by means of TD combined with simultaneous boundary variations by an application of the speed method [245] to Fréchet differentiable shape functionals. Therefore, the general set of optimality conditions is established for a class of shape optimization problems in more general setting compared to the classical theory [83, 245]. To deal with various types of domain modification we introduce the following general notation for different types of variations of shape functionals and of solutions to partial differential equations:

Shape derivative is used in order to determine the variations of solutions to boundary value problems resulting from the boundary variations of geometrical domains. In particular, first the Fréchet differentiability of shape functionals is established, and then the speed method is applied to determine the shape derivative. We refer to [245] and the recent book [83] for general description of the speed method and the related results on Fréchet differentiability of shape functionals.

Topological derivative accounts for variations of shape functionals resulting from the emerging of one or several small holes or cavities in the interior of the geometrical domain.

Domain differential introduced here, unifies the influence on shape functionals of boundary variations and, at the same time, of the nucleation of internal holes or cavities.

Let us recall briefly the definitions used in the shape calculus. In the formulae given below $J(\Omega)$ is an integral functional depending on the solution of the boundary value problem (BVP) defined in Ω. The elliptic systems with the so–called polynomial property [199] are well suited boundary value problems for our analysis. We introduce the mapping $T_\tau : \Omega \to \Omega_\tau$ associated with the vector field $V \in C^1(0, \delta; C_0^2(\mathbb{R}^2, \mathbb{R}^2))$ supported on a small neighbourhood of $\Gamma = \partial\Omega$. The domain Ω_τ is defined by [245]

$$\Omega_\tau = \{x = x(\tau, X) \ : \ X \in \Omega\}$$

where $x = x(\tau, X)$ denotes the solution to the system

$$\frac{dx}{d\tau}(\tau) = V(\tau, x), \qquad x(0) = X.$$

Then the shape derivative of $J(\Omega)$ at Ω^* and in the direction V is given by

$$\mathscr{S}J(\Omega^*; V) = \lim_{\tau \to 0} \frac{1}{\tau} \left[J(T_\tau(\Omega^*)) - J(\Omega^*) \right]. \tag{5.1}$$

Now, let us create a small hole $B_\varepsilon(x) = \{y \in \mathbb{R}^2 : \|x - y\| < \varepsilon\}$ at the point x in the interior of the domain Ω, and let us prescribe the Neumann homogeneous conditions on its boundary. The meaning of the Neumann condition should be made precise for general elliptic systems. According to [238], the topological derivative $\mathscr{T}(\Omega^*; x)$ of

$J(\Omega)$ at $\Omega^* \subset \mathbb{R}^2$ is a function depending on the center x of the small hole and in two dimensions is given by the following limit, if the limit exists,

$$\mathcal{T}(\Omega^*; x) = \lim_{\varepsilon \to 0+} \frac{J(\Omega^* \setminus \overline{B_\varepsilon(x)}) - J(\Omega^*)}{|B_\varepsilon(x)|}. \tag{5.2}$$

Remark 5.1 The following formula [240] is useful in applications, since it uses only the first order shape derivatives of the shape functional $J(\Omega)$ in order to evaluate the topological derivatives of $J(\Omega)$ at $\Omega^* \subset \mathbb{R}^d, d \geq 2$.

$$\mathcal{T}(\Omega^*; x) = \lim_{\varepsilon \downarrow 0} \frac{dJ(\Omega^* \setminus \overline{B_\varepsilon(x)})}{d(|B_\varepsilon(x)|)}, \tag{5.3}$$

where $|B_\varepsilon(x)|$ is the n-dimensional measure of the ball $B_\varepsilon(x)$.

At the end of this chapter we present an example with the optimal solution known explicitly. On the other hand all the computations can be carried out here analytically, in particular for the topological derivatives. The optimal shape in the form of a disk corresponds to a fixed topology, i.e. when only simply connected domains are admissible. When the requirement on the fixed topology is relaxed and still the symmetry of admissible domains is required, the disk remains optimal under the volume constraints. Finally, for the admissible domains which are no more symmetric, the optimality conditions including the topological derivative are not satisfied. Therefore, a topology variation in a precisely indicated region of the disk improves the value of the cost functional. It shows that the domain differential allows us to distinguish all the cases listed above. Therefore the optimality conditions we establish are more precise compared to the classical boundary variation technique. The similar results can be expected for the optimization problems with respect to the geometrical domain involving eigenvalues.

5.2 Model Problem

We are going to analyze a class of shape optimization problems for the Laplace equation. Therefore we need the precise results on asymptotic expansions of solutions to the equation with respect to small parameter $\varepsilon > 0$ which defines the size of geometrical singularities. We consider in this chapter a model problem, therefore the arguments are elementary, and we provide the proofs for the convenience of the reader. In the case of general systems of elliptic equations we refer the reader to [143, 185, 186, 199, 200] for the related results on asymptotic expansions of solutions to elliptic systems with respect to geometrical singularities, and to the paper [205] for asymptotic expansions of arbitrary shape functionals for the second order elliptic systems with constant coefficients.

Let us denote by $C(x_0; \varepsilon)$ a circle $C(x_0; \varepsilon) = \{x : \|x - x_0\| = \varepsilon\}$ and by $P(x_0; \varepsilon, R)$ a ring $P(x_0; \varepsilon, R) = \{x : \varepsilon < \|x - x_0\| < R\}$. If $x_0 = 0$, we simplify notation to $C(\varepsilon)$ and $P(\varepsilon, R)$. We formulate a BVP

$$\begin{cases} \Delta w_\varepsilon = 0 \quad \text{in } P(\varepsilon, R) , \\ w_\varepsilon = 0 \quad \text{on } C(R) , \\ \partial_n w_\varepsilon = h_\varepsilon \text{ on } C(\varepsilon) . \end{cases} \tag{5.4}$$

Now we consider the bounded domain Ω, with boundary consisting of a finite number of smooth arcs, see Fig. 5.1. We define a BVP in Ω:

$$\begin{cases} \Delta u_0 = f \quad \text{in } \Omega , \\ u_0 = g_D \text{ on } \Gamma_D , \\ \partial_n u_0 = g_N \text{ on } \Gamma_N , \end{cases} \tag{5.5}$$

where $\Gamma_N \cup \Gamma_D = \Gamma = \partial\Omega$, $f \in H^m(\Omega)$, and $g_D \in H^{m+3/2}(\Gamma_D)$, $g_N \in H^{m+1/2}(\Gamma_N)$, where $m \geq 0$ and the inclusions are understood in the sense that they hold for every arc in Γ_D, Γ_N, we refer the reader to [204] for details. Under these assumptions, the following theorem holds [204].

Theorem 5.1 *Let $p^1 \dots p^k$ be vertices of Γ (endpoints of smooth arcs),*

$$\Omega_\delta = \Omega \setminus \bigcup_{i=1}^{k} \overline{B_\delta(p^i)}, \qquad \Gamma_\delta = \Gamma \setminus \bigcup_{i=1}^{k} \overline{B_\delta(p^i)}.$$

Then for a given, sufficiently small $\delta > 0$, the solution of (5.5) *satisfies the estimate*

$$\|u_0\|_{H^{m+2}(\Omega_\delta)} \leq \Lambda(\delta) \left(\|f\|_{H^m(\Omega_{\delta/2})} + \|g_D\|_{H^{m+3/2}(\Gamma_{\delta/2})} + \|g_N\|_{H^{m+1/2}(\Gamma_{\delta/2})} \right) .$$

We can replace $\Omega_{\delta/2}$, $\Gamma_{\delta/2}$ by Ω, Γ in the right hand side of the above inequality, provided the norms in the right hand side over the specific sets are finite.

For the convenience of the reader we provide a proof of the asymptotic expansions which are used in the present chapter. Let us recall, that in the polar coordinate system,

$$\Delta w = w_{rr} + \frac{1}{r} w_r + \frac{1}{r^2} w_{\phi\phi}, \tag{5.6}$$

and if w is radially symmetric,

$$\Delta w = w_{rr} + \frac{1}{r} w_r . \tag{5.7}$$

Let $R_0 < R_1$ and $P(R_0, R_1)$ be a ring

$$P(R_0, R_1) = \{x \in \mathbb{R}^2 : R_0 < \|x\| < R_1\}.$$

Define also the circle $C(R)$

$$C(R) = \{x \in \mathbb{R}^2 : \|x\| = R\}.$$

Assume, that $0 < \varepsilon < \frac{1}{2}R_0$ and $R_0 < R$. We define the boundary value problems, parameterized by ε:

$$\begin{cases} \Delta w_\varepsilon = 0 & \text{in } P(\varepsilon, R), \\ w_\varepsilon = 0 & \text{on } C(R), \\ \partial_n w_\varepsilon = -\varepsilon h_\varepsilon & \text{on } C(\varepsilon). \end{cases} \tag{5.8}$$

Furthermore, we assume, that the function h_ε is continuous on $C(\varepsilon)$ and bounded,

$$\|h_\varepsilon\|_{L^2(C(\varepsilon))} \le \Lambda_1,$$

uniformly with respect to ε.

Lemma 5.1 *The function w_ε satisfies the following conditions: (i) on any curve $\Gamma \subset P(R_0, R)$ it may be expressed as*

$$w_\varepsilon = \varepsilon^2 g_\varepsilon,$$

where $|g_\varepsilon| \le \Lambda_2(\Lambda_1)$ on Γ and (ii) on $C(\varepsilon)$ it satisfies inequality

$$|w_\varepsilon| \le \Lambda_3(\Lambda_1)\varepsilon^2 |\ln \varepsilon|.$$

Proof The function h_ε has the Fourier series expansion

$$h_\varepsilon = c_\varepsilon + \sum_{k=1}^{\infty} (a_{\varepsilon,k} \sin k\phi + b_{\varepsilon,k} \cos k\phi),$$

and in addition

$$c_\varepsilon^2 + \sum_{k=1}^{\infty} (a_{\varepsilon,k}^2 + b_{\varepsilon,k}^2) \le \Lambda_1^2. \tag{5.9}$$

We determine the solution w_ε of (5.8) also in the form of a series. The first term, corresponding to c_ε, is radially symmetric, and taking into account (5.7), has the following representation

$$w_\varepsilon^0 = A + B \ln r.$$

From the boundary condition

$$\partial_n w_\varepsilon = -\varepsilon c_\varepsilon \quad \text{on} \quad C(\varepsilon),$$

follows that

$$A + B \ln R = 0 \quad \Rightarrow \quad B\frac{1}{\varepsilon} = \varepsilon c_\varepsilon,$$

hence

$$B = \varepsilon^2 c_\varepsilon, \qquad A = -\varepsilon^2 c_\varepsilon \ln R,$$

where $|c_\varepsilon| \le \Lambda_1$. Finally,

$$w_\varepsilon^0 = \varepsilon^2 c_\varepsilon \ln r - \varepsilon^2 c_\varepsilon \ln R. \tag{5.10}$$

Consider now the term corresponding to the boundary condition

$$\partial_n w_\varepsilon^k = -\varepsilon a_{\varepsilon,k} \sin k\phi \quad \text{on} \quad C(\varepsilon),$$

for $k \ge 1$ (cosine term can be treated in the same way). We seek the solution in the form

$$w_\varepsilon^k = v(r) \sin k\phi,$$

and, considering (5.6), get the representation

$$v(r) = Ar^k + B\frac{1}{r^k}.$$

Again, from the boundary conditions follows, that

$$AR^k + B\frac{1}{R^k} = 0,$$
$$kA\varepsilon^{k-1} - kB\frac{1}{\varepsilon^{k+1}} = \varepsilon a_{\varepsilon,k}.$$

Hence

$$w_\varepsilon^k = \frac{\varepsilon^{k+2} a_{\varepsilon,k}}{k(R^{2k} + \varepsilon^{k-1})}\left(r^k - \frac{R^{2k}}{r^k}\right) \sin k\phi, \tag{5.11}$$

where $|a_{\varepsilon,k}| \le \Lambda_1$. Now, substituting $r := \varepsilon$ in (5.10), (5.11) we get

$$|w_\varepsilon^0(\varepsilon)| \le \Lambda_2(\Lambda_1)\varepsilon^2 |\ln \varepsilon|,$$
$$|w_\varepsilon^k(\varepsilon)| \le \Lambda_3(\Lambda_1)\varepsilon^2.$$

The convergence of the series for the solution w_ε for small ε follows immediately from (5.9).

In the rest of this chapter we assume that all the regularity assumptions concerning the data of the problem (5.5) are satisfied.

5.3 Double Asymptotic Expansion

We consider the case of two distinct holes. The finite number of distinct holes can be treated in the same way. We establish the asymptotic expansions of solutions to the Laplace equation with respect to the radii of the holes. This is an extension of the results given [239] in the case of a single hole.

Let us select two different points x_1 and x_2 in the interior of Ω. Next we remove from Ω two balls, $B_{\varepsilon_1}(x_1)$ and $B_{\varepsilon_2}(x_2)$. There always exists $\delta > 0$, such that the set of pairs $(\varepsilon_1, \varepsilon_2)$ satisfying the condition

$$0 < \varepsilon_1, \varepsilon_2 < \frac{1}{2}\max\{\operatorname{dist}(x_1, \partial\Omega_\delta),\ \operatorname{dist}(x_2, \partial\Omega_\delta),\ \frac{1}{2}\|x_2 - x_1\|\ \} \tag{5.12}$$

is nonempty. We denote

$$\Omega_{\varepsilon_1\varepsilon_2} = \Omega \setminus (\overline{B_{\varepsilon_1}(x_1)} \cup \overline{B_{\varepsilon_2}(x_2)}) \tag{5.13}$$

and denote by $u_{\varepsilon_1\varepsilon_2}$ the solution of the following BVP, which is a modification of (5.5) taking into account the existence of holes,

$$\begin{cases} \Delta u_{\varepsilon_1\varepsilon_2} = f & \text{in } \Omega_{\varepsilon_1\varepsilon_2}\ , \\ u_{\varepsilon_1\varepsilon_2} = g_D & \text{on } \Gamma_D\ , \\ \partial_n u_{\varepsilon_1\varepsilon_2} = g_N & \text{on } \Gamma_N\ , \\ \partial_n u_{\varepsilon_1\varepsilon_2} = 0 & \text{on } C(x_1;\varepsilon_1) \cup C(x_2;\varepsilon_2)\ . \end{cases} \tag{5.14}$$

Furthermore, we assume that the number m defining the regularity of the data as specified in Theorem 5.1 satisfies $m \geq 1$.

In view of the condition (5.12) satisfied by the radii $\varepsilon_1, \varepsilon_2$ there exist constants R_0 and R, such that $R_0 > \max\{\varepsilon_1, \varepsilon_2\}$ and

$$\Omega \subset B_R(x_1) \cap B_R(x_2), \quad B_{\varepsilon_1}(x_1) \cap \partial\Omega_\delta = B_{\varepsilon_2}(x_2) \cap \partial\Omega_\delta = \emptyset.$$

All the geometrical objects introduced so far are presented in Fig. 5.1.

In order to formulate the main result of this section, we define the functions s_{ε_1}, s_{ε_2} of the form

$$\begin{aligned} s_{\varepsilon_1}(x) &= \frac{\partial u_0}{\partial x_1}(x_1)\frac{\varepsilon_1^2}{r_1}\cos\theta_1 + \frac{\partial u_0}{\partial x_2}(x_1)\frac{\varepsilon_1^2}{r_1}\sin\theta_1, \\ s_{\varepsilon_2}(x) &= \frac{\partial u_0}{\partial x_1}(x_2)\frac{\varepsilon_2^2}{r_2}\cos\theta_2 + \frac{\partial u_0}{\partial x_2}(x_2)\frac{\varepsilon_2^2}{r_2}\sin\theta_2. \end{aligned}$$

Here u_0 is the solution of (5.5), while $(r_1 = \|x - x_1\|, \theta_1)$ and $(r_2 = \|x - x_2\|, \theta_2)$ are polar coordinate systems around x_1 and x_2 respectively.

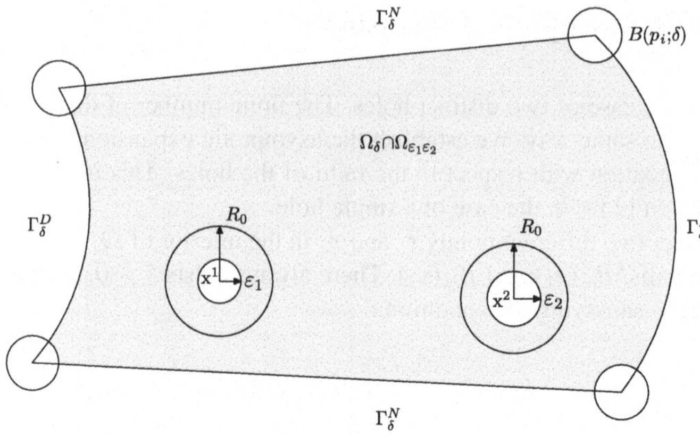

Fig. 5.1 The domains Ω_δ, $\Omega_{\varepsilon_1\varepsilon_2}$ and related geometrical objects

We prove, modifying the argument given in [239] for a single hole, that the following asymptotic expansion is obtained in the case of two holes.

Lemma 5.2 *The solution of* (5.14) *may be expressed as follows*

$$u_{\varepsilon_1\varepsilon_2} = u_0 + s_{\varepsilon_1} + s_{\varepsilon_2} + z_{\varepsilon_1\varepsilon_2},$$

where $z_{\varepsilon_1\varepsilon_2}$ *satisfies on* $C(x_1;\varepsilon_1) \cup C(x_2;\varepsilon_2)$ *the estimates*

$$|z_{\varepsilon_1\varepsilon_2}| \leq \Lambda \left(\varepsilon_1^2 |\log \varepsilon_1| + \varepsilon_2^2 |\log \varepsilon_2|\right),$$
$$\|\nabla z_{\varepsilon_1\varepsilon_2}\| \leq \Lambda(\varepsilon_1 + \varepsilon_2),$$

and in addition, on the set $P(x_1; R_0, R) \cap P(x_2; R_0, R) \cap \Omega$

$$|z_{\varepsilon_1\varepsilon_2}| \;\leq\; \Lambda(\varepsilon_1^2 + \varepsilon_2^2),$$

with $\Lambda = \Lambda(\Omega, \delta, R_0, R)$.

Proof Let us set $\varepsilon = \sqrt{\varepsilon_1^2 + \varepsilon_2^2}$. We shall express $z_{\varepsilon_1\varepsilon_2}$ in the form of the series

$$z_{\varepsilon_1\varepsilon_2} = p^1_{\varepsilon_1} + p^1_{\varepsilon_2} + q^1_\varepsilon + \varepsilon^2(p^2_{\varepsilon_1} + p^2_{\varepsilon_2} + q^2_\varepsilon) + \dots$$

where the consecutive terms are defined by BVP's specified below. For $p^1_{\varepsilon_1}$ we have:

$$\begin{cases} \Delta p^1_{\varepsilon_1} = 0 & \text{in } P(x_1;\varepsilon_1, R)\,, \\ p^1_{\varepsilon_1} = 0 & \text{on } C(x_1; R)\,, \\ \partial_n p^1_{\varepsilon_1} = -\partial_n(u_0 + s_{\varepsilon_1}) & \text{on } C(x_1;\varepsilon_1)\,, \end{cases}$$

and for $p^1_{\varepsilon_2}$:

$$\begin{cases} \Delta p^1_{\varepsilon_2} = 0 & \text{in } P(x_2;\varepsilon_2,R)\,, \\ p^1_{\varepsilon_2} = 0 & \text{on } C(x_2;R)\,, \\ \partial_n p^1_{\varepsilon_2} = -\partial_n(u_0 + s_{\varepsilon_2}) & \text{on } C(x_2;\varepsilon_2)\,, \end{cases}$$

while

$$\begin{cases} \Delta q^1_\varepsilon = 0 & \text{in } \Omega\,, \\ q^1_\varepsilon = -s_{\varepsilon_1} - s_{\varepsilon_2} - p^1_{\varepsilon_1} - p^1_{\varepsilon_2} & \text{on } \Gamma\,. \end{cases}$$

The idea of this construction is as follows: functions $p^1_{\varepsilon_1}$ and $p^1_{\varepsilon_2}$ correct the normal derivative of $z_{\varepsilon_1\varepsilon_2}$ on $C(x_1;\varepsilon_1)$ and $C(x_2;\varepsilon_2)$ respectively. However, $p^1_{\varepsilon_1}$ introduces the discrepancy around $C(x_2;\varepsilon_2)$ and $p^1_{\varepsilon_2}$ introduces the discrepancy around $C(x_1;\varepsilon_1)$. In addition, $p^1_{\varepsilon_1}, p^1_{\varepsilon_2}$ disturb the boundary conditions on Γ. Therefore q^1_ε is introduced to correct the boundary conditions on Γ. Now observe, that due to the regularity of u_0 (see Theorem 5.1), ∇u_0 is continuous on the sets $\overline{B_{R_0}(x_1)}$ and $\overline{B_{R_0}(x_2)}$. Thus, the norm of the gradient $\|\nabla u_0\|$ is bounded on these sets and we have

$$\left|\partial_n(u_0 + s_{\varepsilon_1})\right| \le A\varepsilon_1 \quad \text{on} \quad C(x_1;\varepsilon_1),$$

$$\left|\partial_n(u_0 + s_{\varepsilon_2})\right| \le A\varepsilon_2 \quad \text{on} \quad C(x_2;\varepsilon_2).$$

Therefore, by Lemma 5.1,

$$\begin{aligned} |p^1_{\varepsilon_1}| \le A\varepsilon_1^2|\log\varepsilon_1| & \quad \text{on} \quad C(x_1;\varepsilon_1)\,, \\ |p^1_{\varepsilon_1}| + \|\nabla p^1_{\varepsilon_1}\| \le A\varepsilon_1^2 & \quad \text{on} \quad P(x_1;R_0,R)\,, \\ |p^1_{\varepsilon_2}| \le A\varepsilon_2^2|\log\varepsilon_2| & \quad \text{on} \quad C(x_2;\varepsilon_2)\,, \\ |p^1_{\varepsilon_2}| + \|\nabla p^1_{\varepsilon_2}\| \le A\varepsilon_2^2 & \quad \text{on} \quad P(x_2;R_0,R)\,. \end{aligned}$$

Hence, $|q^1_\varepsilon| \le A\varepsilon^2$ on Γ, and also

$$\|\nabla q^1_\varepsilon\| \le A\varepsilon^2 \quad \text{on} \quad P(x_1;\varepsilon_1,R_0) \cup P(x_2;\varepsilon_2,R_0).$$

In the next step we compute the corrections of higher order. As we have seen above, the Neumann boundary conditions on $C(x_1;\varepsilon_1)$ are perturbed by the functions $p^1_{\varepsilon_2}$ and q^1_ε. Therefore, the next set of functions $p^2_{\varepsilon_1}, p^2_{\varepsilon_2}, q^2_\varepsilon$, is defined by solutions to the following BVP's. For $p^2_{\varepsilon_1}$:

$$\begin{cases} \Delta p^2_{\varepsilon_1} = 0 & \text{in } P(x_1;\varepsilon_1,R)\,, \\ p^2_{\varepsilon_1} = 0 & \text{on } C(x_1;R)\,, \\ \partial_n p^2_{\varepsilon_1} = -\varepsilon^{-2}\partial_n(q^1_\varepsilon + p^1_{\varepsilon_2}) & \text{on } C(x_1;\varepsilon_1)\,, \end{cases}$$

and for $p^2_{\varepsilon_2}$:

$$\begin{cases} \Delta p_{\varepsilon_2}^2 = 0 & \text{in } P(x_2; \varepsilon_2, R) , \\ p_{\varepsilon_2}^2 = 0 & \text{on } C(x_2; R) , \\ \partial_n p_{\varepsilon_2}^2 = -\varepsilon^{-2} \partial_n (q_\varepsilon^1 + p_{\varepsilon_1}^1) & \text{on } C(x_2; \varepsilon_2) , \end{cases}$$

while q_ε^2 satisfies

$$\begin{cases} \Delta q_\varepsilon^2 = 0 & \text{in } \Omega , \\ q_\varepsilon^2 = -\varepsilon^{-2}(p_{\varepsilon_1}^2 + p_{\varepsilon_2}^2) & \text{on } \Gamma . \end{cases}$$

In addition, from properties of $p_{\varepsilon_1}^1, p_{\varepsilon_2}^1, q_\varepsilon^1$ it follows that

$$\begin{aligned} \left|\partial_n (q_\varepsilon^1 + p_{\varepsilon_2}^1)\right| &\le \Lambda \varepsilon^2 && \text{on} \quad C(x_1; \varepsilon_1) , \\ \left|\partial_n (q_\varepsilon^1 + p_{\varepsilon_1}^1)\right| &\le \Lambda \varepsilon^2 && \text{on} \quad C(x_2; \varepsilon_2) . \end{aligned}$$

By Lemma 5.1 the terms $\varepsilon^2(p_{\varepsilon_1}^2 + p_{\varepsilon_2}^2 + q_\varepsilon^2)$ constitute the correction of the order $\varepsilon^3 |\log \varepsilon|$. Finally, we use again Lemma 5.1 to obtain the estimates for $p_{\varepsilon_1}^1 + p_{\varepsilon_2}^1 + q_\varepsilon^1$ and therefore, for $z_{\varepsilon_1 \varepsilon_2}$.

5.4 Topological Differential with Respect to Multiple Holes

We shall restrict our analysis to the case of two holes, since the generalization to the case of a finite number of holes can be performed in the same way. Let us define the domain functionals I_u and I_g as follows

$$I_u(\varepsilon_1, \varepsilon_2) = \int_{\Omega_{\varepsilon_1 \varepsilon_2}} F(u_{\varepsilon_1 \varepsilon_2})\, dx, \tag{5.15}$$

$$I_g(\varepsilon_1, \varepsilon_2) = \int_{\Omega_{\varepsilon_1 \varepsilon_2}} |\nabla u_{\varepsilon_1 \varepsilon_2}|^{2p}\, dx, \tag{5.16}$$

$p = 1, 2$, where F is a C^2 function, $u_{\varepsilon_1 \varepsilon_2}$ solves (5.14) and $\Omega_{\varepsilon_1 \varepsilon_2}$ is defined in (5.13), retaining the notation already introduced in previous sections. Our main result is given in the form of the following theorem.

Theorem 5.2 *Assume that $\varepsilon_1, \varepsilon_2 > 0$ are small enough. Then the following representation is obtained for the topological variations of the shape functionals:*

$$I_u(\varepsilon_1, \varepsilon_2) = I_u(0, 0) + \mathcal{T}_u(x_1)\, |B_{\varepsilon_1}(x_1)| + \mathcal{T}_u(x_2)\, |B_{\varepsilon_2}(x_2)| + o(\varepsilon^2), \tag{5.17}$$

$$I_g(\varepsilon_1, \varepsilon_2) = I_g(0, 0) + \mathcal{T}_g(x_1)\, |B_{\varepsilon_1}(x_1)| + \mathcal{T}_g(x_2)\, |B_{\varepsilon_2}(x_2)| + o(\varepsilon^2). \tag{5.18}$$

We omit here the dependence of the topological derivative on the domain Ω, see (5.2), because Ω is fixed in all subsequent considerations. The formulae (5.17), (5.18) define an object, e.g., for the functional I_u,

$$\mathcal{T}_u(\Omega; x_1, x_2) = [\mathcal{T}_u(\Omega; x_1), \mathcal{T}_u(\Omega; x_2)] ,$$

which, by analogy to **topological derivative**, we shall call a **topological gradient**. The name is chosen because of its similarity to the gradient of an ordinary multivariable function consisting of partial derivatives. The topological gradient allows us to express variations of shape functionals in terms of volumes of balls centered at x_1, x_2 (in general at finite number of points).

Proof We compute the directional derivative of $I_u(\varepsilon_1, \varepsilon_2)$. Before proceeding further, let us recall that in case of a single hole the formula for $\mathcal{T}_u(x_0)$ has been derived in [238] and takes the form

$$\mathcal{T}_u(x_0) = -\left[F(u_0) + f w_0 + 2\nabla u_0 \cdot \nabla w_0\right]_{x=x_0} , \tag{5.19}$$

where w_0 is the adjoint functions which satisfies (5.22). Let us take $c_1, c_2 > 0$, $c_1^2 + c_2^2 = 1$, and the radii of balls in the form $\varepsilon_1 = c_1\varepsilon$, $\varepsilon_2 = c_2\varepsilon$. Then $\varepsilon_1^2 + \varepsilon_2^2 = \varepsilon^2$ and $|B_{\varepsilon_1}(x_1)| + |B_{\varepsilon_2}(x_2)| = \pi\varepsilon^2$. By varying the parameter ε we change the boundaries of both holes simultaneously. Finally, we introduce the spaces

$$\begin{aligned} H_D^1(\Omega) &= \{\phi \in H^1(\Omega) \ : \ \phi = 0 \quad \text{on} \quad \Gamma_D\}, \\ H_g^1(\Omega) &= \{\phi \in H^1(\Omega) \ : \ \phi = g_D \quad \text{on} \quad \Gamma_D\}. \end{aligned}$$

In the first step we evaluate the derivative

$$\begin{aligned} \frac{d}{d\varepsilon} I_u(\varepsilon_1, \varepsilon_2) = \int_{\Omega_{\varepsilon_1\varepsilon_2}} & F_u(u_{\varepsilon_1\varepsilon_2}) u'_{\varepsilon_1\varepsilon_2} \, dx \\ & - c_1 \int_{C(x_1;\varepsilon_1)} F(u_{\varepsilon_1\varepsilon_2}) \, ds - c_2 \int_{C(x_2;\varepsilon_2)} F(u_{\varepsilon_1\varepsilon_2}) \, ds, \end{aligned} \tag{5.20}$$

where $u'_{\varepsilon_1\varepsilon_2}$ is the shape derivative of $u_{\varepsilon_1\varepsilon_2}$ with respect to the change of radii of both circles, we refer to [245] for the details. By $d/d\varepsilon$ we mean the Eulerian semiderivative of the shape functional $I_u(\varepsilon_1, \varepsilon_2)$ in the direction of the specific vector field V equal to the exterior unit normal field $-n$ on $C(x_1; \varepsilon_1)$ and $C(x_2; \varepsilon_2)$. We apply here in fact the Hadamard formula [245] which says the in the case of differentiable shape functionals the shape derivatives actually depend on the normal component of the velocity field on the moving boundary. We know, that

$$u'_{\varepsilon_1\varepsilon_2} = \dot{u}_{\varepsilon_1\varepsilon_2} - \nabla u_{\varepsilon_1\varepsilon_2} \cdot V,$$

where V is an appropriate vector field introduced above by an extension of the normal field $-n$ to the small neighbourhoods of $C(x_1; \varepsilon_1)$ and $C(x_2; \varepsilon_2)$, and $\dot{u}$ denotes the material derivative. Now, observe that we may take smooth V, which vanishes outside Ω_δ, i.e. the only region where $\nabla u_{\varepsilon_1\varepsilon_2}$ actually may have singularities. Hence $u'_{\varepsilon_1\varepsilon_2}$ enjoys the same regularity as $\dot{u}_{\varepsilon_1\varepsilon_2}$, in particular $u'_{\varepsilon_1\varepsilon_2} \in H_D^1(\Omega_{\varepsilon_1\varepsilon_2})$. The shape derivative $u'_{\varepsilon_1\varepsilon_2} \in H^1_D(\Omega_{\varepsilon_1\varepsilon_2})$ satisfies for all $\phi \in H^1_D(\Omega_{\varepsilon_1\varepsilon_2})$ the following integral identity

$$\begin{aligned}\int_{\Omega_{\varepsilon_1\varepsilon_2}} \nabla u'_{\varepsilon_1\varepsilon_2} \cdot \nabla\phi\, dx - c_1 \int_{C(x_1;\varepsilon_1)} \nabla u'_{\varepsilon_1\varepsilon_2} \cdot \nabla\phi\, ds - c_2 \int_{C(x_2;\varepsilon_2)} \nabla u'_{\varepsilon_1\varepsilon_2} \cdot \nabla\phi\, ds \\ = c_1 \int_{C(x_1;\varepsilon_1)} f\phi\, ds + c_2 \int_{C(x_2;\varepsilon_2)} f\phi\, ds. \qquad (5.21)\end{aligned}$$

Now we introduce the adjoint variables w_0 and $w_{\varepsilon_1\varepsilon_2}$ defined by the following variational identities: Find $w_0 \in H^1_D(\Omega)$ such that

$$-\int_{\Omega} \nabla w_0 \cdot \nabla\phi, dx = \int_{\Omega} F_u(u_0)\phi\, dx \quad \text{for all} \quad \phi \in H^1_D(\Omega), \qquad (5.22)$$

and find $w_{\varepsilon_1\varepsilon_2} \in H^1_D(\Omega_{\varepsilon_1\varepsilon_2})$ such that

$$-\int_{\Omega_{\varepsilon_1\varepsilon_2}} \nabla w_{\varepsilon_1\varepsilon_2} \cdot \nabla\phi\, dx = \int_{\Omega_{\varepsilon_1\varepsilon_2}} F_u(u_{\varepsilon_1\varepsilon_2})\phi\, dx \quad \text{for all} \quad \phi \in H^1_D(\Omega_{\varepsilon_1\varepsilon_2}). \qquad (5.23)$$

Taking into account the regularity of $u'_{\varepsilon_1\varepsilon_2}$, we make cross–substitutions in (5.21) and (5.23) which leads to

$$\begin{aligned}\int_{\Omega_{\varepsilon_1\varepsilon_2}} F_u(u_{\varepsilon_1\varepsilon_2})u'_{\varepsilon_1\varepsilon_2}\, dx =& - c_1 \int_{C(x_1;\varepsilon_1)} \left[f w_{\varepsilon_1\varepsilon_2} + \nabla u_{\varepsilon_1\varepsilon_2} \cdot \nabla w_{\varepsilon_1\varepsilon_2}\right] ds \\ &- c_2 \int_{C(x_2;\varepsilon_2)} \left[f w_{\varepsilon_1\varepsilon_2} + \nabla u_{\varepsilon_1\varepsilon_2} \cdot \nabla w_{\varepsilon_1\varepsilon_2}\right] ds. \qquad (5.24)\end{aligned}$$

As a result, from (5.20) we obtain

$$\begin{aligned}\frac{d}{d\varepsilon} I_u(\varepsilon_1,\varepsilon_2) =& - c_1 \int_{C(x_1;\varepsilon_1)} \left[F(u_{\varepsilon_1\varepsilon_2}) + f w_{\varepsilon_1\varepsilon_2} + \nabla u_{\varepsilon_1\varepsilon_2} \cdot \nabla w_{\varepsilon_1\varepsilon_2}\right] ds \\ &- c_2 \int_{C(x_2;\varepsilon_2)} \left[F(u_{\varepsilon_1\varepsilon_2}) + f w_{\varepsilon_1\varepsilon_2} + \nabla u_{\varepsilon_1\varepsilon_2} \cdot \nabla w_{\varepsilon_1\varepsilon_2}\right] ds \qquad (5.25)\end{aligned}$$

In the next step we observe that the derivative with respect to ε^2 can be expressed by the derivative with respect to ε in the following way,

$$\frac{dI_u(\varepsilon_1,\varepsilon_2)}{d(\pi\varepsilon^2)} = \frac{1}{2\pi\varepsilon}\frac{d}{d\varepsilon} I_u(\varepsilon_1,\varepsilon_2). \qquad (5.26)$$

Such an observation is very useful for applications, since we avoid any evaluation of the second order derivatives of shape functionals for the specific two dimensional problem. In addition, it can be easily proved in the same way as it has been done in

[238], by an application of the results given by [204], for the case of single hole, that $w_{\varepsilon_1\varepsilon_2}$ has the same sort of expansion as $u_{\varepsilon_1\varepsilon_2}$,

$$w_{\varepsilon_1\varepsilon_2} = w_0 + s_{\varepsilon_1}(w_0) + s_{\varepsilon_2}(w_0) + z_{\varepsilon_1\varepsilon_2}(w_0),$$

with $z_{\varepsilon_1\varepsilon_2}(w_0)$ satisfying the same type of estimates, namely on the set $C(x_1;\varepsilon_1) \cup C(x_2;\varepsilon_2)$

$$|z_{\varepsilon_1\varepsilon_2}(w_0)| \le \Lambda\left(\varepsilon_1^2|\log\varepsilon_1| + \varepsilon_2^2|\log\varepsilon_2|\right),$$
$$\|\nabla z_{\varepsilon_1\varepsilon_2}(w_0)\| \le \Lambda(\varepsilon_1 + \varepsilon_2),$$

and on the set $P(x_1; R_0, R) \cap P(x_2; R_0, R) \cap \Omega$

$$|z_{\varepsilon_1\varepsilon_2}(w_0)| \le \Lambda(\varepsilon_1^2 + \varepsilon_2^2).$$

Taking into account both expansions for $u_{\varepsilon_1\varepsilon_2}$ and $w_{\varepsilon_1\varepsilon_2}$ we may compute the limit

$$\lim_{\varepsilon\to 0+} \frac{(-1)}{2\pi\varepsilon} c_1 \int_{C(x_1;\varepsilon_1)} \left[F(u_{\varepsilon_1\varepsilon_2}) + f w_{\varepsilon_1\varepsilon_2} + \nabla u_{\varepsilon_1\varepsilon_2}\cdot\nabla w_{\varepsilon_1\varepsilon_2}\right] ds$$
$$= -c_1^2\left[F(u_0) + f w_0 + 2\nabla u_0\cdot\nabla w_0\right]_{x=x_1} = c_1^2\, \mathcal{T}_u(x_1), \tag{5.27}$$

and similarly for the integral on $C(x_2;\varepsilon_2)$. This gives for the increment of I_u the expression

$$dI_u = \pi\varepsilon^2\left[c_1^2\, \mathcal{T}_u(x_1) + c_1^2\, \mathcal{T}_u(x_2)\right].$$

But $\pi\varepsilon^2 c_1^2 = |B_{\varepsilon_1}(x_1)|$ and $\pi\varepsilon^2 c_2^2 = |B_{\varepsilon_2}(x_2)|$, therefore the expansion (5.17) for I_u follows. The expansion for the functional I_g can be obtained in the same way.

5.5 Dependence of Solutions on Boundary Variations

We describe briefly the speed method in shape sensitivity analysis, referring the reader to [245] for details. The domains with moving boundaries are constructed in such a way that allows us to obtain expansions of solutions to BVP's in such domains with respect to the boundary perturbations.

Let us consider the BVP similar to the problem discussed in previous sections, with the difference that the Neumann part of the boundary Γ of the domain Ω is split into two parts: a fixed part Γ_N, with nonhomogeneous Neumann conditions, and the remaining moving part Γ^V, i.e. the part subjected to the boundary variations, with the homogeneous Neumann condition.

$$\begin{cases} \Delta u_0 = f & \text{in } \Omega \,, \\ u_0 = g_D & \text{on } \Gamma_D \,, \\ \partial_n u_0 = g_N & \text{on } \Gamma_N \,, \\ \partial_n u_0 = 0 & \text{on } \Gamma^V \,. \end{cases} \tag{5.28}$$

In addition, we assume that Γ^V consists of a single smooth arc and the parameter m describing the regularity of the data in Theorem 5.1 satisfies now stronger condition $m \geq 2$.

The evolution of Γ^V with respect to the parameter τ is defined by the transformation $T(\tau, \cdot) : \Omega \mapsto \Omega_\tau$ (see [245]), depending on the vector field Θ, of the form

$$T(\tau, x) \;=\; x + \tau\Theta(x), \qquad x \in B_R \supset \Omega. \tag{5.29}$$

We assume that the vector field Θ is smooth enough, $\Theta(x) \in \left[C^4(\overline{B_R})\right]^2$, and

$$\|\Theta\|_{\left[C^4(\overline{B_R})\right]^2} \leq C_\theta.$$

Moreover, let $\delta > 0$ such a constant that the field $\Theta(x)$ vanishes on $U = \Omega \setminus (\Gamma_N \cup \Gamma_D + B_\delta)$, i.e. on the neighbourhood of the fixed part of the boundary.

These assumptions imply several properties of the transformation $T(\tau, \cdot)$. Namely there exists a constant $\tau_0 > 0$ such that for $\tau < \tau_0$ we have:

1. $T(\tau, \cdot)$ is invertible in B_R;
2. $T(\tau, x) = x$ on U, and $T(\tau, B_R \setminus U) = B_R \setminus U$;
3. $T^{-1}(\tau, y) = y - \tau K_1(y, \tau)$, where

$$\|K_1(y, \tau)\|_{\left[C^3(\overline{B_R})\right]^2} \leq A C_\theta.$$

The third property requires short justification. Let $y = T(\tau, x)$. Then

$$\begin{aligned} y - \tau\Theta(y) &= x + \tau\Theta(x) - \tau\Theta(x + \tau\Theta(x)) \\ &= x + \tau\Theta(x) - \tau\Theta(x) - \tau^2 D\Theta(y') \cdot \Theta(x), \end{aligned}$$

where $y' = x - \eta\tau\Theta(x)$ and $0 < \eta < 1$. Hence we have the representation $x = y - \tau K_1(y, \tau)$, where $K_1 = \Theta(y) - \tau D\Theta(y') \cdot \Theta(x)$ is bounded in C^3-norm.

Let us now fix $\delta > 0$ and let $\Theta(x)$ be a vector field satisfying all conditions listed above. We consider in the domain $\Omega_\tau = T(\tau, \Omega)$ the following BVP parameterized by $\tau \geq 0$,

$$\begin{cases} \Delta u_\tau = f & \text{in } \Omega_\tau \,, \\ u_\tau = g_D & \text{on } \Gamma_D \,, \\ \partial_n u_\tau = g_N & \text{on } \Gamma_N \,, \\ \partial_n u_\tau = 0 & \text{on } \Gamma_\tau^V = T(\tau, \Gamma^V) \,. \end{cases} \tag{5.30}$$

In the above system all the differentiations are done with respect to the variable $y \in \Omega_\tau$.

Lemma 5.3 *For $\tau \geq 0$, τ small enough, the solution of* (5.30) *can be expressed as a function of the parameter τ as follows:*

$$u_\tau = u_0 + \tau z,$$

where

$$\|z\|_{H^3(\Omega_\delta)} \leq \Lambda(\Omega, C_\theta),$$

and we denote $C_\theta = \|\Theta\|_{C^3(\overline{B_R})}$.

Proof After the change of variables defined by the transformation $y = T(\tau, x)$ the system (5.30) is transported in a standard way [245] to the fixed domain Ω, and the resulting system takes on the following form

$$\begin{cases} \operatorname{div}\left[J(T)(D_yT^{-1}D_y^{-T})\nabla u_\tau\right] = J(T)f(T(\tau,x)) & \text{in } \Omega\,, \\ u_\tau = g_D & \text{on } \Gamma_D\,, \\ \partial_n u_\tau = g_N & \text{on } \Gamma_N\,, \\ n^T(D_yT^{-1}D_y^{-T})\nabla u_\tau = 0 & \text{on } \Gamma^V\,. \end{cases} \tag{5.31}$$

The coefficients appearing in the above system may be expressed in the form suitable for further computations. Namely

$$D_yT^{-1}(\tau,y) = I - \tau D_y K_1(y,\tau),$$
$$J(T) = det\,(I + \tau D\Theta(x)) = 1 + \tau k_1(\tau,x)$$

where

$$\|k_1(\cdot,\tau)\|_{C^3(\overline{B_R})} \leq \Lambda C_\theta.$$

Therefore we have the representations

$$D_yT^{-1}D_y^{-T} = (I - \tau D_y K_1)(I - \tau D_y K_1^T) = I + \tau K_2(\tau,x),$$
$$J(T)(D_yT^{-1}D_y^{-T}) = I + \tau K_3(\tau,x),$$
$$J(T)f(T(\tau,x)) = f(x) + \tau f_1(x,\tau),$$

in which

$$\|K_2(\tau,\cdot)\|_{[C^2(\overline{B_R})]^4},\quad \|K_3(\tau,\cdot)\|_{[C^2(\overline{B_R})]^4} \leq \Lambda C_\theta \quad \text{and} \quad \|f_1(\tau,\cdot)\|_{H^1(\Omega)} \leq \Lambda C_\theta.$$

As a result the transformed system (5.31) takes on the form

$$\begin{cases} \operatorname{div}[(I+\tau K_3)\nabla u_\tau] = f + \tau f_1 & \text{in } \Omega\ , \\ u_\tau = g_D & \text{on } \Gamma_D\ , \\ \partial_n u_\tau = g_N & \text{on } \Gamma_N\ , \\ n^T(I+\tau K_2)\nabla u_\tau = 0 & \text{on } \Gamma^V\ . \end{cases} \tag{5.32}$$

It is important to observe, that all the functions K_1, K_2, K_3, k_1 vanish on U. Now we apply the method of matched asymptotics to the system (5.31), assuming that the solution u_τ can be expanded in the form of the series with respect to the parameter τ,

$$u_\tau = \sum_{k=0}^{\infty} \tau^k u_k(\tau, x). \tag{5.33}$$

Substituting (5.33) into (5.32) leads formally to the system

$$\begin{cases} \sum_{k=0}^{\infty} \tau^k \Delta u_k + \sum_{k=0}^{\infty} \tau^{k+1} \operatorname{div}(K_3 \nabla u_k) = f + \tau f_1 & \text{in } \Omega\ , \\ \sum_{k=0}^{\infty} \tau^k u_k = g_D & \text{on } \Gamma_D\ , \\ \sum_{k=0}^{\infty} \tau^k \partial_n u_k = g_N & \text{on } \Gamma_N\ , \\ \sum_{k=0}^{\infty} \tau^k \partial_n u_k + \sum_{k=0}^{\infty} \tau^{k+1} n^T K_2 \nabla u_k = 0 & \text{on } \Gamma^V\ . \end{cases} \tag{5.34}$$

In a standard way, comparing the terms with the powers of τ of the same order, we obtain the sequence of BVP's for subsequent functions u_k, $k = 0, 1, \ldots$. Thus u_0 satisfies (5.28), for $k = 1$ we have

$$\begin{cases} \Delta u_1 = f_1 - \operatorname{div}(K_3 \nabla u_0) & \text{in } \Omega\ , \\ u_1 = 0 & \text{on } \Gamma_D\ , \\ \partial_n u_1 = 0 & \text{on } \Gamma_N\ , \\ \partial_n u_1 = -n^T K_2 \nabla u_0 & \text{on } \Gamma^V\ , \end{cases}$$

and for $k > 1$

$$\begin{cases} \Delta u_k = -\operatorname{div}(K_3 \nabla u_{k-1}) & \text{in } \Omega\ , \\ u_k = 0 & \text{on } \Gamma_D\ , \\ \partial_n u_k = 0 & \text{on } \Gamma_N\ , \\ \partial_n u_k = -n^T K_2 \nabla u_{k-1} & \text{on } \Gamma^V\ . \end{cases}$$

By the trace theorem

$$\|\operatorname{div}(K_3 \nabla u_{k-1})\|_{H^1(\Omega_\delta)} + \|n^T K_2 \nabla u_{k-1}\|_{H^{5/2}(\Gamma_\delta^V)} \le \Lambda_3\, C_\Theta \|u_{k-1}\|_{H^3(\Omega_\delta)}.$$

But, since K_3, K_2 vanish on U, we have

$$\|\operatorname{div}(K_3 \nabla u_{k-1})\|_{H^1(\Omega_\delta)} = \|\operatorname{div}(K_3 \nabla u_{k-1})\|_{H^1(\Omega_{\delta/2})},$$

$$\|n^T K_2 \nabla u_{k-1}\|_{H^{5/2}(\Gamma_\delta^V)} = \|n^T K_2 \nabla u_{k-1}\|_{H^{5/2}(\Gamma_{\delta/2}^V)}$$

and therefore

$$\|\mathrm{div}\,(K_3\nabla u_{k-1})\|_{H^1(\Omega_{\delta/2})} + \|n^T K_2 \nabla u_{k-1}\|_{H^{5/2}(\Gamma^V_{\delta/2})} \le \Lambda_3\, C_\theta \|u_{k-1}\|_{H^3(\Omega_\delta)}.$$

Thus, due to Theorem 5.1, we are able to obtain the recursive bounds for the consecutive solutions u_k:

$$\text{for } k=1: \quad \|u_1\|_{H^3(\Omega_\delta)} \le \Lambda(\|f_1\|_{H^1(\Omega_{\delta/2})} + \Lambda C_\theta \|u_0\|_{H^3(\Omega_{\delta/2})}) \le \Lambda C_\theta (1 + \|u_0\|_{H^3(\Omega)}),$$
$$\text{for } k>1: \quad \|u_k\|_{H^3(\Omega_\delta)} \le \Lambda C_\theta \|u_{k-1}\|_{H^3(\Omega_\delta)}.$$

As a result there exists a constant $\tau_1 > 0$ such that for $\tau < \min[\tau_0, \tau_1]$ the series (5.33) is convergent in $H^3(\Omega_\delta)$. The last step consists in verification that u_τ is indeed a solution. We take the truncated series

$$u_\tau^N = \sum_{k=0}^{N} \tau^k u_k(\tau, x)$$

and substitute it into the initial BVP (5.32), obtaining

$$\begin{cases} \sum_{k=0}^N \tau^k \Delta u_k + \sum_{k=0}^N \tau^{k+1} \mathrm{div}\,(K_3 \nabla u_k) = f + \tau f_1 + \tau^{N+1} \mathrm{div}\,(K_3 \nabla u_\tau^N) & \text{in } \Omega\,, \\ u_\tau^N = g_D & \text{on } \Gamma_D\,, \\ \partial_n u_\tau^N = g_N & \text{on } \Gamma_N\,, \\ n^T(I + \tau K_2)\nabla u_\tau^N = \tau^{N+1} n^T K_2 \nabla u_\tau^N & \text{on } \Gamma^V\,. \end{cases}$$

Hence for the remainder of the series we have the following estimate

$$\|u_\tau^N - u_\tau\|_{H^3(\Omega_\delta)} \le \Lambda \tau^{N+1}$$

which completes the proof.

Remark 5.2 The same result can be obtained for the nonhomogeneous Neumann condition or for the nonhomogeneous Dirichlet condition on the moving part of the boundary. The only difference with the proof of Lemma 5.3 is that the functions g and h are transported to the fixed domain and then expanded with respect to the parameter τ. The construction of the asymptotic expansions is performed in the same way as for the right hand side of the equation in the proof of Lemma 5.3. Therefore, the method of matched asymptotic expansions with respect to τ is applicable for the problems with nonhomogeneous boundary conditions of Neumann or Dirichlet types.

5.6 Simultaneous Topology and Shape Modification

In this section we shall investigate the variation of the integral goal functional resulting from the nucleation of an internal hole and from the boundary variation. We assume that the volume $|\Omega|$ of the geometrical domain is preserved by such perturbations. Let all the requirements concerning the domain and the field Θ specified in previous sections hold. We assume in addition that for a given $\delta > 0$ the support of the vector field Θ is contained in the tubular neighborhood $\Gamma^V + B_{\delta/2}$. This condition ensures that under our assumptions on $\Theta(x)$ the following properties of the mapping $T(\tau, \cdot) : \mathbb{R}^2 \to \mathbb{R}^2$ are obtained for $|\tau|$ small enough:

- $T(\tau, \cdot)$ is a bijection of $\Gamma^V + B_\delta$ onto itself;
- $T(\tau, \cdot)$ is the identity mapping on the set $\Omega \setminus (\Gamma^V + B_\delta)$.

For a given $x_0 \in \Omega$, and $\delta > 0$ such that $x_0 \in \Omega \setminus (\Gamma^V + B_\delta)$, we select a field $\Theta(x)$ and assume $|\tau|$ to be small enough. We denote by $\Omega_{\varepsilon\tau}$ the domain $\Omega_\tau \setminus \overline{B_\varepsilon(x_0)}$.

Finally, $u_{\varepsilon\tau}$ is a solution of the following BVP defined on $\Omega_{\varepsilon\tau}$, with the regularity of the data specified by $m \geq 2$.

$$\begin{cases} \Delta u_{\varepsilon\tau} = f & \text{in } \Omega_{\varepsilon\tau} , \\ u_{\varepsilon\tau} = g_D & \text{on } \Gamma_D , \\ \partial_n u_{\varepsilon\tau} = g_N & \text{on } \Gamma_N , \\ \partial_n u_{\varepsilon\tau} = 0 & \text{on } C(x_0; \varepsilon) \cup \Gamma_\tau^V . \end{cases} \tag{5.35}$$

The form of $\Omega_{\varepsilon\tau}$ and the geometry of the problem under considerations is shown in Fig. 5.2. Similarly, as in previous sections, we consider shape functionals of the form

$$I_u(\eta, \tau) = \int_{\Omega_{\varepsilon\tau}} F(u_{\varepsilon\tau})\, dx, \qquad I_g(\eta, \tau) = \int_{\Omega_{\varepsilon\tau}} \|\nabla u_{\varepsilon\tau}\|^{2p}\, dx , \tag{5.36}$$

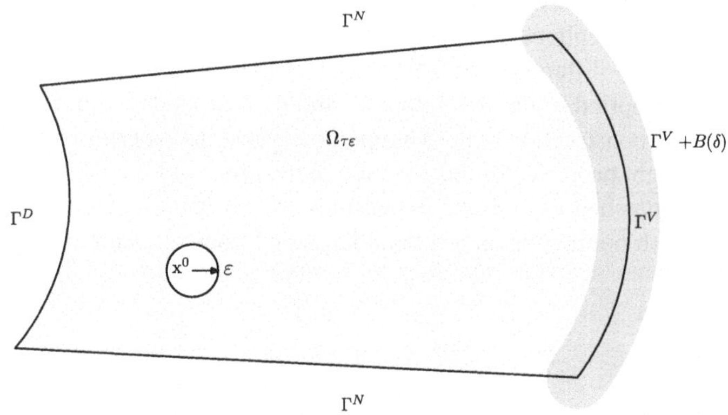

Fig. 5.2 The configuration of the internal hole and variable boundary

where we denote $\eta = \pi\varepsilon^2$. Finally, according to (5.1), we denote by $\mathscr{S}I_u(\Omega;\Theta)$ and $\mathscr{S}I_g(\Omega;\Theta)$ the shape derivatives of the functionals $I_u(\eta,\tau)$, $I_g(\eta,\tau)$, with respect to the variation of Γ^V, taken at Ω, i.e. for $\eta = 0$, $\tau = 0$. It is well known [245] that these derivatives are given by the following formulae,

$$\mathscr{S}I_u(\Omega;\Theta) = \int_{\Gamma^V} [F(u_{00}) + \nabla u_{00}\cdot\nabla w_{00}]\,(\Theta\cdot n)\,ds, \tag{5.37}$$

$$\mathscr{S}I_g(\Omega;\Theta) = \int_{\Gamma^V} \left[\|\nabla u_{00}\|^{2p} + \nabla u_{00}\cdot\nabla v_{00}\right](\Theta\cdot n)\,ds. \tag{5.38}$$

Furthermore, these derivatives depend only on a normal component $\Theta_n = \Theta\cdot n$ of Θ on Γ^V. The adjoint variable w_{00} satisfies the Eq. (5.22). The adjoint variable v_{00} is a solution of the following adjoint equation depending on the gradient of the functional I_g, find $v_{00}\in H^1_D(\Omega)$ such that

$$-\int_\Omega \nabla v_{00}\cdot\nabla\phi, dx = \int_\Omega 2p|\nabla u_{00}|^{2p-2}(\nabla u_{00}\cdot\nabla\phi)\,dx \quad \text{for all} \quad \phi\in H^1_D(\Omega).$$

We present the form of the variations of shape functionals resulting from modification of the geometrical domain by boundary variations and by nucleation of small holes.

Theorem 5.3 *If all the assumptions stated above are satisfied, the functionals $I_u(\eta,\tau)$ and $I_g(\eta,\tau)$ have the representation (with $\eta = \pi\varepsilon^2$):*

$$I_u(\eta,\tau) = I_u(0,0) + \eta\mathscr{T}_u(\Omega;x_0) + \tau\mathscr{S}I_u(\Omega;\Theta) + o(\eta) + o(\tau),$$
$$I_g(\eta,\tau) = I_g(0,0) + \eta\mathscr{T}_u(\Omega;x_0) + \tau\mathscr{S}I_g(\Omega;\Theta) + o(\eta) + o(\tau).$$

Proof The proof is given in the case of I_u, since for the second shape functional I_g the same argument applies. Let us consider the expression

$$I_u(\eta,\tau) - I_u(0,0) = I_u(\eta,\tau) - I_u(0,\tau) + I_u(0,\tau) - I_u(0,0). \tag{5.39}$$

From the results on the shape sensitivity analysis [245] it follows that

$$I_u(0,\tau) - I_u(0,0) = \tau\mathscr{S}I_u(\Omega;\Theta) + o(\tau).$$

It remains to analyse the first difference on the right hand side of (5.39). By an application of Theorem 5.2 for a single hole we have

$$I_u(\eta,\tau) - I_u(0,\tau) = \eta\mathscr{T}_u(\Omega_\tau;x_0) + o(\eta), \tag{5.40}$$

and by Lemma 5.2 the term $o(\eta)$ is uniform with respect to τ, for $|\tau|$ small enough. But we have the explicit form of the topological derivative in Ω_τ,

$$\mathcal{T}_u(\Omega_\tau; x_0) = -[F(u_{0\tau}) + f w_{0\tau} + 2\nabla u_{0\tau} \cdot \nabla w_{0\tau}]_{x=x_0} .$$

Using again Lemma 5.2, we have

$$\begin{aligned} 2|u_{0\tau}(x_0) - u_{00}(x_0)| &\le \Lambda\tau, & \|\nabla u_{0\tau}(x_0) - \nabla u_{00}(x_0)\| &\le \Lambda\tau, \\ |w_{0\tau}(x_0) - w_{00}(x_0)| &\le \Lambda\tau, & \|\nabla w_{0\tau}(x_0) - \nabla w_{00}(x_0)\| &\le \Lambda\tau. \end{aligned}$$

Substituting these estimates into (5.40) leads to

$$I_u(\eta, \tau) - I_u(0, \tau) = \eta\mathcal{T}_u(\Omega; x_0) + o(\eta) + O(\eta\tau),$$

and the required result for I_u follows. The case of I_g may be treated in an analogous way.

Remark 5.3 Let us note that using the method proposed in this chapter we can define the **domain differential** denoted by $\mathcal{D}J(\Omega; \Theta, x)$ of an arbitrary shape functional $J(\Omega)$,

$$\mathcal{D}J(\Omega; \Theta, x_0)(\varepsilon, \tau) = |B_\varepsilon(x_0)|\ \mathcal{T}(\Omega; x_0) + \tau\ \mathcal{S}J(\Omega; \Theta) .$$

Such a differential provides complete characterization of the variation of $J(\Omega)$ with respect to the variations of Ω, taking into account both the shape and topology changes.

From Theorem 5.3 it follows by standard arguments that an optimal domain Ω^* satisfies condition

$$\mathcal{D}J(\Omega^*; \Theta, x_0)(\varepsilon, \tau) = |B_\varepsilon(x_0)|\ \mathcal{T}(\Omega^*; x_0) + \tau\ \mathcal{S}J(\Omega; \Theta) \ge 0$$

for all admissible (ε, τ) and all admissible vector fields Θ in an appropriate tangent set. In the case e.g., of volume constraints the above formula leads to the following necessary optimality conditions

$$\mathcal{T}(\Omega^*; x_0) \ge 0 \quad \text{in } \Omega \quad \text{and} \quad \mathcal{S}J(\Omega; \Theta) \ge 0 \quad \forall\Theta . \tag{5.41}$$

The latter inequality follows since under the volume constraints $|\Omega_{\varepsilon\tau}| = |\Omega|$ we have the relation [236, 245],

$$\pi\varepsilon^2 = \tau \int_{\Gamma^V} \theta_n d\Gamma + o(\tau)$$

which can be neglected for the admissible tangent directions of Θ. The form of some tangent sets in L^∞ can be found in [41].

Fig. 5.3 Domain $\Omega \subset \mathbb{R}^2$

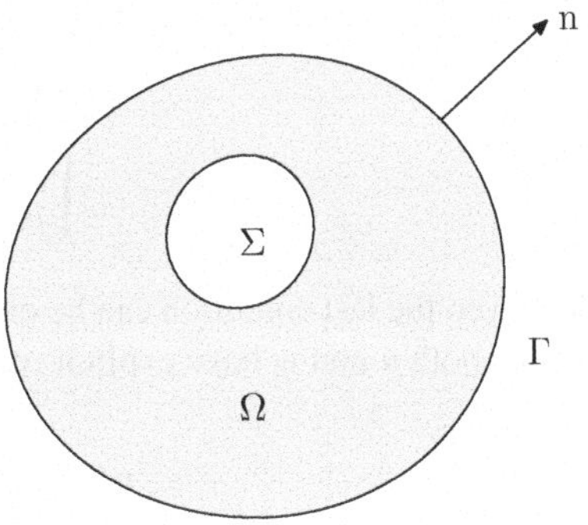

5.7 Analytical Example

In the following example it is shown that using the method proposed in this chapter we can verify that the shape optimal in the framework of classical theory can be improved using the topology variations. For the problem under considerations the second order sufficient optimality conditions has been established in [44]. Let us consider the shape functional

$$J(\Omega) = \int_{\Omega} u^2 \, dx \quad \to \quad \max, \tag{5.42}$$

where $\Omega \subset \mathbb{R}^2$ is a ring with exterior boundary Γ and interior boundary Σ, as shown in Fig. 5.3. The function u satisfies a BVP

$$\begin{cases} -\Delta u = 1 \text{ in } \Omega, \\ u = 0 \text{ on } \Gamma, \\ \partial_n u = 0 \text{ on } \Sigma. \end{cases} \tag{5.43}$$

We impose the volume constraints $|\Omega| = \pi$ for admissible domains. This example is motivated by optimum design problems concerning the elastic bars in torsion.

First, we assume that $\Sigma = \emptyset$, i.e. only the Dirichlet boundary conditions are prescribed on the external boundary Γ. Then, it is known [44] that the unit ball $\Omega^* = B_1$ maximizes $J(\Omega)$ over the family of simply connected domains with the Dirichlet boundary condition prescribed on the external boundary. We may easily check that the domain Ω^* is a critical point of the integral shape functional $J(\Omega)$. Let us introduce the mapping $T_\tau : \Omega \to \Omega_\tau$ associated with the vector field V supported in a small neighbourhood of $\Gamma^* = \partial\Omega^*$. In this section the speed vector field is denoted by V instead of Θ. If the volume of Ω_τ is to be preserved, $|\Omega_\tau| = |\Omega|$, then we have

$$\int_{\Gamma^*} (V \cdot n) \, ds = 0. \tag{5.44}$$

The adjoint equation for the problem under considerations has the form

$$\begin{cases} \Delta w = 2u & \text{in } \Omega^*, \\ w = 0 & \text{on } \Gamma^*, \\ \partial_n w = 0 & \text{on } \Sigma, \end{cases} \tag{5.45}$$

however, the last condition can be omitted since $\Sigma = \emptyset$. For such a simple geometry of Ω^* both u and w have explicit representations in polar coordinates (r, θ):

$$u(x) = \frac{1}{4}(1 - r^2),$$
$$w(x) = \frac{1}{32}r^4 - \frac{1}{8}r^2 + \frac{3}{32}.$$

Therefore the shape derivative of $J(\Omega)$ at Ω^* and in the direction V is given by

$$\begin{aligned} \mathscr{S}J(\Omega^*; V) &= \int_{\Gamma^*} \partial_n w \, \partial_n u \, (V \cdot n) \, ds \\ &= \frac{1}{16} \int_{\Gamma^*} (V \cdot n) \, ds = 0 \qquad \text{due to (5.44).} \end{aligned} \tag{5.46}$$

Thus Ω^* is a critical point.

Let us consider the change of the topology by nucleation of a small circular hole B_ε inside of the domain $\Omega^* = B_1$, with the Neumann part of the boundary $\Sigma_\varepsilon = \partial B_\varepsilon$. In order to preserve the volume we must move the exterior boundary Γ^* and expand Ω^*, thus we introduce for $\varepsilon > 0$

$$\Omega_\varepsilon = B_{\sqrt{1+\varepsilon^2}} \setminus \overline{B_\varepsilon}, \qquad \Gamma_\varepsilon = \partial B_{\sqrt{1+\varepsilon^2}},$$

and denote by u_ε the solution to (5.43) in Ω_ε. Again, we have the explicit expressions

$$u_\varepsilon(x) = u(x) + \frac{1}{4}\varepsilon^2 + \frac{1}{2}\varepsilon^2 \log \frac{r}{\sqrt{1+\varepsilon^2}},$$
$$J(\Omega_\varepsilon) = \pi \left[\frac{1}{48} - \frac{3}{32}\varepsilon^2 + \frac{1}{4}\varepsilon^6 A^2 + \frac{1}{8}\varepsilon^6 A - \frac{1}{4}\varepsilon^4 A + \frac{1}{16}\varepsilon^4 \right],$$

with

$$A = \log \frac{\varepsilon}{\sqrt{1+\varepsilon^2}}.$$

It is easy to see that both u_ε and ∇u_ε converge pointwise to u and ∇u in B_1, with the removable singularity at $x = 0$. Moreover $\varepsilon = 0$, what corresponds to $\Omega_\varepsilon = \Omega^* = B_1$, still gives the maximum of the functional $J(\Omega_\varepsilon)$ defined above, see Fig. 5.4.

Using explicit expression for $J(\Omega_\varepsilon)$ we may directly compute the limit

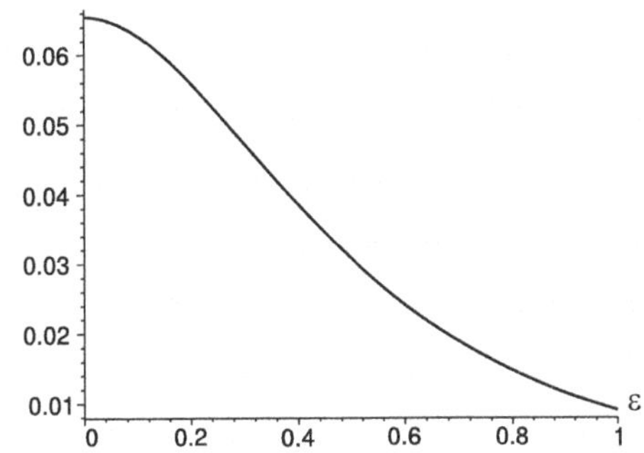

Fig. 5.4 The dependence the goal functional $J(\Omega_\varepsilon)$ as a function of the radius of internal hole ε

$$\lim_{\varepsilon\to 0+} \frac{dJ(\Omega_\varepsilon)}{d(\pi\varepsilon^2)} = \lim_{\varepsilon\to 0+} \frac{1}{2\pi\varepsilon}\frac{dJ(\Omega_\varepsilon)}{d\varepsilon} = -\frac{3}{32}. \tag{5.47}$$

The negative value indicates, as expected, that a hole of a sufficiently small area with the center at $x = 0$ decreases the value of the functional. Now we compute the same limit using the domain differential, with the shape and topological parts. According to [238] and (5.19), the topological derivative is given by

$$\mathscr{T}(\Omega^*; 0) = -\left[(u(0))^2 - w(0) + 2\nabla u(0)\cdot\nabla w(0)\right] = -\frac{5}{32}.$$

In order to preserve the volume, we set $(V \cdot n) = 1$ in the shape derivative on Γ^* and select τ such that $2\pi\tau = \pi\varepsilon^2$. Then, in view of (5.46),

$$\mathscr{S}J(\Omega^*; V) = \frac{1}{16}\int_{\Gamma^*}(V\cdot n)\,ds = \frac{1}{16}2\pi.$$

Hence

$$\begin{aligned}\mathscr{D}J(\Omega^*; V, 0)(\varepsilon,\tau) &= \pi\varepsilon^2\mathscr{T}(\Omega^*; 0) + \tau\mathscr{S}J(\Omega^*; V)\\ &= \left(-\frac{5}{32}+\frac{1}{16}\right)\pi\varepsilon^2 = -\frac{3}{32}\pi\varepsilon^2,\end{aligned}$$

in agreement with (5.47). This computation confirms the formula given by Theorem 5.3 for the simple case.

The natural question is, if without the assumption on the radial symmetry, i.e., with a small hole at $r \neq 0$, the value of the shape functional can possibly be improved. We compute the domain differential

$$\mathscr{D}J(\Omega^*; V, x) = \left[\frac{1}{16} - (u(r))^2 + w(r) - 2\frac{\partial u}{\partial r}(r)\frac{\partial w}{\partial r}(r)\right]\pi\varepsilon^2 = \pi\varepsilon^2 f(r).$$

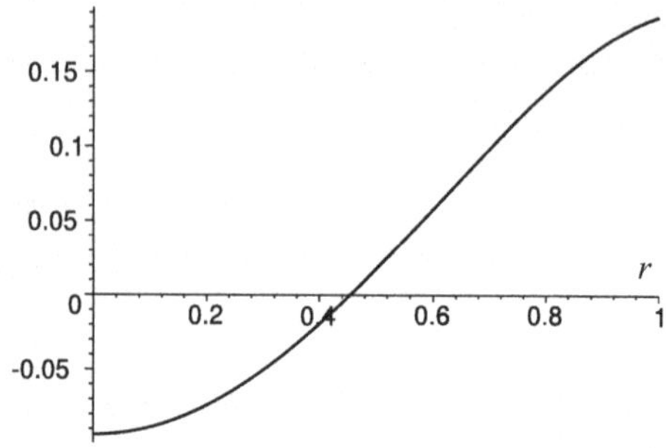

Fig. 5.5 The dependence of the function $f(r)$ with respect to the location of the hole r, showing possibility of increasing the functional

The graph of the function $f(r)$ is given in Fig. 5.5. We have $f(r) > 0$ for the location $r > r_0 \approx 0.45$, what means that the nucleation of a small hole near the edge increases $J(\Omega)$. Thus looking at the domain differential we conclude that the unit ball *is not* an optimal domain for the problem (5.42)–(5.43) with volume constraint, if we change the topology and admit an arbitrarily located $\Sigma \neq \emptyset$.

Chapter 6
A Gradient-Type Method and Applications

In classical optimization the first family of numerical methods is based on the linear approximation of the functional to be minimized. The same idea is used in the topological derivative method. The functional is approximated by the first order expansion with respect to the small parameter. The parameter governs the size of singular domain perturbation which we call the topological perturbation. The numerical method converges to a local solution of the necessary optimality conditions, as it is shown by the numerical experiments. The framework of topological sensitivity analysis in singularly perturbed geometrical domains allows to propose the levelset-type method for topology optimization. The asymptotic expansion of a given shape functional is constructed with respect to a small parameter that measures the size of small holes, cavities, inclusions, source-terms and cracks. This approximation technique is new in the shape sensitivity analysis. The topological derivative generalizes the classical shape derivative from the domain boundary to its interior for classes of admissible domains in two and three spatial dimensions. The case of one spatial dimension is also possible for such expansions but it is more delicate issue. Nowadays, the concept of topological derivative is a new and powerful tool for solving shape-topology optimization problems for elliptic boundary value problems. There are now applications of the topological derivative method in many different fields of engineering and physics, such as shape and topology optimization, inverse problems, image processing, multi-scale material design and mechanical modeling including damage and fracture evolution phenomena. In this chapter a topology optimization algorithm based on first order topological derivatives is presented. The appropriate level-set domain representation method is employed within the iterations in order to design an optimal shape-topology local solution. The algorithm is successfully used for numerical solution of a wide class of shape-topology optimization problems.

A. A. Novotny et al., *Applications of the Topological Derivative Method*, Studies in Systems, Decision and Control 188,
https://doi.org/10.1007/978-3-030-05432-8_6

6.1 Preliminaries

The topological derivative represents the first term of the asymptotic expansion of a given shape functional with respect to a small parameter which measures the size of singular domain perturbations, such as holes, inclusions, source-terms and cracks. This relatively new concept has been successfully applied in many different fields, including shape and topology optimization, inverse problems, image processing, multiscale material design and mechanical modeling involving damage and fracture evolution phenomena.

It is worth mentioning that topological derivative is defined through a limit passage when the small parameter governing the size of the topological perturbation goes to zero. Therefore, it can be used as a steepest-descent direction in an optimization process, just as in any method based on the gradient of the cost functional. In particular, in this chapter a topology optimization algorithm based on first order topological derivatives together with a level-set domain representation method is presented [25]. Finally, these ideas are used to solve a wide class of topology optimization problems.

We restrict ourselves to the case in which the domain is topologically perturbed by the nucleation of a small inclusion. This allows for working in a fixed computational domain $\mathscr{D} \subset \mathbb{R}^d$, where a weak material phase is used to mimic voids. This simple strategy avoids the use of a complicated algorithm specifically designed to deal with nucleation of holes in a computational domain. Let us consider that the hold-all domain $\mathscr{D}$ is split into two subdomains, $\Omega \subset \mathscr{D}$ and its complement $\mathscr{D} \setminus \Omega$. We assume that there is a distributed parameter $\rho : \mathscr{D} \mapsto \{1, \rho_0\}$ defined as

$$\rho(x) := \begin{cases} 1 & \text{if } x \in \Omega, \\ \rho_0 & \text{if } x \in \mathscr{D} \setminus \Omega. \end{cases} \tag{6.1}$$

with $0 < \rho_0 \ll 1$. Let us introduce a shape functional $\Omega \mapsto J(\Omega)$. The topology optimization problem we are dealing with is as follows:

$$\underset{\Omega \subset \mathscr{D}}{\text{Minimize}} \ J(\Omega) \ , \tag{6.2}$$

which can be solved by using the topological derivative concept. Actually, a hole $\omega_\varepsilon(\widehat{x})$ is introduced inside $\mathscr{D}$. Then, the region occupied by $\omega_\varepsilon(\widehat{x})$ is filled by an inclusion with material properties different from the background. The material properties are characterized by a piecewise constant function γ_ε of the form

$$\gamma_\varepsilon(x) := \begin{cases} 1 & \text{if } x \in \mathscr{D} \setminus \overline{\omega_\varepsilon} \ , \\ \gamma(x) & \text{if } x \in \omega_\varepsilon \ , \end{cases} \tag{6.3}$$

where the contrast γ is defined as

$$\gamma(x) = \begin{cases} \rho_0 & \text{if } x \in \Omega, \\ \rho_0^{-1} & \text{if } x \in \mathscr{D} \setminus \Omega, \end{cases} \tag{6.4}$$

which induces a level-set domain representation method. We explain these ideas in more detail in Sect. 6.2. The topology optimization problem we are dealing with is presented in Sect. 6.3, together with a wide class of applications. In particular, Sect. 6.3.1 deals with several structure topology optimization problems. In Sect. 6.3.2 a fluid flow channel design problem is presented. Section 6.3.3 is dedicated to the synthesis of materials in a multiscale framework. Some additional applications found in the current literature are discussed in Sect. 6.3.4. Finally, the chapter ends in Sect. 6.4 with a discussion concerning perspectives for future development, together with a list of open problems.

6.2 First Order Topology Design Algorithm

In this section a topology optimization algorithm based on first order topological derivatives together with a level-set domain representation method is presented. It has been proposed in [25] and consists basically in achieving a local optimality condition for the minimization problem (6.2), given in terms of the topological derivative and a level-set function. In particular, the domain $\Omega \subset \mathscr{D}$ and the complement $\mathscr{D} \setminus \Omega$ are characterized by a level-set function Ψ:

$$\Omega = \{x \in \mathscr{D} : \Psi(x) < 0\} \quad \text{and} \quad \mathscr{D} \setminus \Omega = \{x \in \mathscr{D} : \Psi(x) > 0\}, \tag{6.5}$$

where Ψ vanishes on the interface between Ω and $\mathscr{D} \setminus \Omega$. A local sufficient optimality condition for Problem (6.2), under a class of domain perturbations given by ball-shaped inclusions denoted by $B_\varepsilon(x)$, can be stated as [24]

$$\mathscr{T}(x) > 0 \quad \forall x \in \mathscr{D}, \tag{6.6}$$

where $\mathscr{T}(x)$ is the topological derivative of the shape functional $J(\Omega)$ at $x \in \mathscr{D}$ and $B_\varepsilon(x)$ is a ball of radius ε centered at $x \in \mathscr{D}$, as shown in Fig. 6.1. Therefore, let us define the quantity

$$g(x) := \begin{cases} -\mathscr{T}(x) \text{ if } \Psi(x) < 0, \\ +\mathscr{T}(x) \text{ if } \Psi(x) > 0, \end{cases} \tag{6.7}$$

which allows rewriting the condition (6.6) in the following equivalent form:

$$\begin{cases} g(x) < 0 \text{ if } \Psi(x) < 0, \\ g(x) > 0 \text{ if } \Psi(x) > 0. \end{cases} \tag{6.8}$$

We observe that (6.8) is satisfied, where the quantity g coincides with the level-set function Ψ up to a strictly positive factor, namely $\exists\, \tau > 0 : g = \tau \Psi$, or equivalently

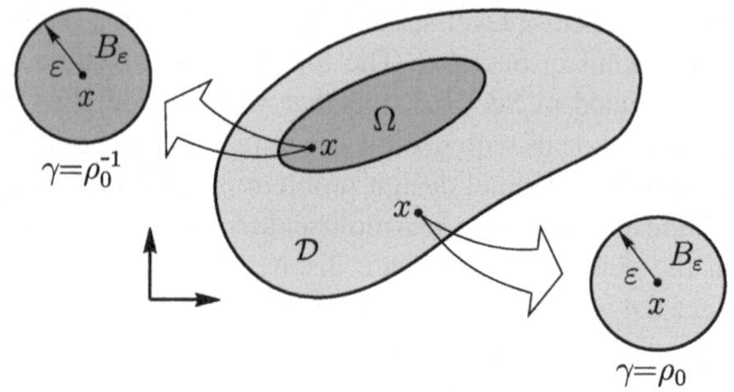

Fig. 6.1 Nucleation of a ball-shaped inclusion $B_\varepsilon(x)$

$$\theta := \arccos\left[\frac{\langle g, \Psi\rangle_{L^2(\mathscr{D})}}{\|g\|_{L^2(\mathscr{D})}\|\Psi\|_{L^2(\mathscr{D})}}\right] = 0, \tag{6.9}$$

which will be used as the optimality condition in the topology design algorithm, where θ is the angle in $L^2(\mathscr{D})$ between the functions g and Ψ.

Let us now explain the algorithm. We start by choosing an initial level-set function Ψ_0. In a generic iteration n, we compute the function g_n associated with the level-set function Ψ_n. Thus, the new level-set function Ψ_{n+1} is updated according to the following linear combination between the functions g_n and Ψ_n:

$$\begin{aligned}
&\Psi_0 : \|\Psi_0\|_{L^2(\mathscr{D})} = 1,\\
&\Psi_{n+1} = \frac{1}{\sin\theta_n}\left[\sin((1-k)\theta_n)\Psi_n + \sin(k\theta_n)\frac{g_n}{\|g_n\|_{L^2(\mathscr{D})}}\right] \forall n \in \mathbb{N},
\end{aligned} \tag{6.10}$$

where θ_n is the angle between g_n and Ψ_n, and k is a step size determined by a line-search performed in order to decrease the value of the objective function $J(\Omega_n)$, with Ω_n used to denote the domain associated with Ψ_n. The process ends when the condition $\theta_n \leq \varepsilon_\theta$ is satisfied at some iteration, where ε_θ is a given small numerical tolerance. Since we have chosen $\Psi_0 : \|\Psi_0\|_{L^2(\mathscr{D})} = 1$, by construction $\Psi_{n+1} : \|\Psi_{n+1}\|_{L^2(\mathscr{D})} = 1$ $\forall n \in \mathbb{N}$. If at some iteration n the line-search step size k is found to be smaller then a given numerical tolerance $\varepsilon_k > 0$ and the optimality condition is not satisfied, namely $\theta_n > \varepsilon_\theta$, then a mesh refinement of the hold-all domain $\mathscr{D}$ is carried out and the iterative process is continued. The resulting first order topology design algorithm is summarized in pseudo-code format in Algorithm 1. For further applications of this algorithm, see for instance [10, 29, 31, 33, 107, 176, 177, 227, 248].

In the context of topological-derivative-based topology optimization methods, the algorithms available in the literature usually combine topological derivatives with shape derivatives or level-set methods [6, 63, 92, 221], leading to a two-stage topology/shape optimization procedure. More precisely, new holes are nucleated according to the topological derivative, while standard tools in shape optimization are used to move the new boundaries. In contrast, Algorithm 1 is based on the optimality condition (6.6) in terms of the topological derivative and a level-set function, leading

Algorithm 1: The topology design algorithm

input : $\mathscr{D}$, Ψ_0, ε_k, ε_θ;
output: the optimal topology $\Omega^\star$;

1 $n \leftarrow 0$;
2 $\Omega_n \leftarrow \Psi_n$;
3 compute the shape functional $J(\Omega_n)$;
4 compute the associated topological derivative $\mathscr{T}(x)$;
5 compute g_n and θ_n according to (6.7) and (6.9);
6 $\Psi_{\text{old}} \leftarrow \Psi_n$; $J_{\text{old}} \leftarrow J(\Omega_n)$; $J_{\text{new}} \leftarrow 1 + J_{\text{old}}$; $k \leftarrow 1$;
7 **while** $J_{\text{new}} > J_{\text{old}}$ **do**
8   compute Ψ_{new} according to (6.10);
9     $\Psi_n \leftarrow \Psi_{new}$;
10   execute lines 2 and 3;
11     $J_{\text{new}} \leftarrow J(\Omega_n)$;
12     $k \leftarrow k/2$;
13 **end while**
14 **if** $k < \varepsilon_k$ **then**
15   try a mesh refinement;
16     $\Psi_{n+1} \leftarrow \Psi_n$; $n \leftarrow n + 1$;
17   **go to** line 2;
18 **else if** $\theta_n > \varepsilon_\theta$ **then**
19     $\Psi_{n+1} \leftarrow \Psi_n$; $n \leftarrow n + 1$;
20   **go to** line 2;
21 **else**
22   **return** $\Omega^\star \leftarrow \Psi_n$;
23   **stop;**
24 **end if**

to a very simple and quite efficient one-stage algorithm driven by the topological derivative only. However, how to efficiently use the topological derivative in the context of topology optimization deserves further investigation. See Sect. 6.4 for an account of some open problems.

6.3 Shape and Topology Optimization

The topological derivative has been specifically designed to deal with shape and topology optimization. In contrast to traditional topology optimization methods, the topological derivative formulation does not require a material model concept based on intermediary densities, so that interpolation schemes are unnecessary. These features are crucial in a wide range of applications, since the limitations arising from material model procedures are here naturally avoided. In addition, topological derivative has the advantage of providing an analytical form for the topological sensitivity which allows to obtain the optimal design in a few iterations or even in just one shot.

Therefore, the resulting topology optimization algorithms are remarkably efficient and of simple computational implementation, since it features only a minimal number of user-defined algorithmic parameters. In this section, the fist order topology design Algorithm 1 is applied in the context of shape and topology optimization. See also related works [6, 52, 63, 112, 150, 167, 215–217, 252].

6.3.1 Structural Topology Design

This section deals with structural topology optimization problems [4, 47, 93, 219]. We start with the classical problem of structural compliance minimization under volume constraint. The next two examples concern volume minimization under stress constraints in the context of structural optimization and design of compliant mechanisms, respectively. Finally, we present an example associated with structural topology optimization under loading uncertainties.

Let us introduce the hold-all domain $\mathscr{D} \subset \mathbb{R}^d$, $d \geq 2$, with Lipschitz boundary $\Gamma := \partial\mathscr{D}$. The elasticity boundary value problem we are dealing with is stated as follows: Find a displacement vector field u such that

$$\begin{cases} -\mathrm{div}\sigma(u) = 0 & \text{in } \mathscr{D} , \\ \sigma(u) = \mathbb{C}(\nabla u)^s , & \\ u = 0 & \text{on } \Gamma_D , \\ \sigma(u)n = \overline{q} & \text{on } \Gamma_N . \end{cases} \tag{6.11}$$

In the above system, $(\nabla u)^s$ is the symmetric part of the gradient of u, and $\mathbb{C}$ is the fourth order elasticity tensor, which can be written in terms of the Lamé coefficients μ and λ as follows:

$$\mathbb{C} = 2\mu\mathbb{I} + \lambda(\mathrm{I} \otimes \mathrm{I}) , \tag{6.12}$$

with I and $\mathbb{I}$ used to denote the second and fourth order identity tensors, respectively. In addition, $\Gamma = \Gamma_D \cup \Gamma_N$ with $\Gamma_D \cap \Gamma_N = \varnothing$, where Γ_D and Γ_N are Dirichlet and Neumann boundaries, respectively. Thus $\overline{q}$ is a Neumann data on Γ_N, assumed to be smooth enough. The strain energy stored in the elastic body is minimized under a volume constraint. Therefore, the topology optimization problem we are dealing with consists in finding a subdomain $\Omega \subset \mathscr{D}$ that solves the following minimization problem:

$$\underset{\Omega \subset \mathscr{D}}{\mathrm{Minimize}}\ \mathscr{F}_\Omega(u) = \mathscr{J}(u) + \beta|\Omega| , \tag{6.13}$$

where β is a fixed multiplier used to impose a volume constraint in Ω of the form $|\Omega| \leq M$ and $\mathscr{J}(u)$ is the energy shape functional, that is,

$$\mathcal{J}(u) = \frac{1}{2}\int_{\mathcal{D}} \sigma(u) \cdot (\nabla u)^s \,. \tag{6.14}$$

The vector function u is a solution to the elasticity boundary value problem (6.11). In particular, by fixing different values of β we get different volume fractions at the end of the iterative process. For more sophisticated topological-derivative-based methods with volume constraint we refer the reader to [66], for instance.

We consider the elasticity boundary value problem into three spatial dimensions, $\mathcal{D} \subset \mathbb{R}^3$. The topological perturbation we are dealing with consists in nucleating a small spherical cavity $\omega_\varepsilon(x) = B_\varepsilon(x)$, with the homogeneous Neumann boundary condition on ∂B_ε. This means that the cavity has a free boundary, so that it represents a void embedded within the elastic body $\mathcal{D}$. In this case, the topological derivative of $\mathcal{F}_\Omega(u)$ is given by the sum

$$\mathcal{T}(x) = \mathcal{T}_E(x) + \beta\mathcal{T}_V(x) \quad \forall x \in \Omega \,. \tag{6.15}$$

The last term $\mathcal{T}_V(x)$ represents the topological derivative of the volume, which is trivially given by

$$\mathcal{T}_V(x) = \begin{cases} -1 \text{ if } x \in \Omega, \\ +1 \text{ if } x \in \mathcal{D} \setminus \Omega, \end{cases} \tag{6.16}$$

while the topological derivative of the energy $\mathcal{J}(u)$ is known [219, Ch. 8, pp. 213], whose closed formula is given by [105, 171, 217]

$$\mathcal{T}_E(x) = \mathbb{P}\sigma(u(x)) \cdot (\nabla u(x))^s \quad \forall x \in \mathcal{D} \,, \tag{6.17}$$

where $\mathbb{P}$ is the polarization tensor, given in this particular case by the following isotropic fourth order tensor [16]:

$$\mathbb{P} = \frac{3}{4}\frac{1-\nu}{7-5\nu}\left(10\mathbb{I} - \frac{1-5\nu}{1-2\nu}\mathrm{I}\otimes\mathrm{I}\right) \,, \tag{6.18}$$

where E is the Young modulus and ν the Poisson ratio.

In order to briefly explain the significance of topological derivative in shape optimization, we present a benchmark example borrowed from [219]. We use a simple procedure consisting in a successive nucleation of cavities where the topological derivative is most negative. In particular, the topology is identified by the strong material distribution, and inclusions of weak material are used to mimic cavities. In addition, the topological derivative is evaluated at the nodal points of a finite element mesh. Then, we remove the elements that share the node where the topological derivative assumes its most negative values. This procedure is repeated until the topological derivative becomes positive everywhere. For a more elaborate topology design algorithm, the reader may refer to [25]. In particular, let us consider as initial guess a cube simply supported at the bottom under vertical load applied on the top, as shown in Fig. 6.2. The details of the results obtained are shown in Fig. 6.3.

Fig. 6.2 Initial guess and boundary conditions for the trestle design problem

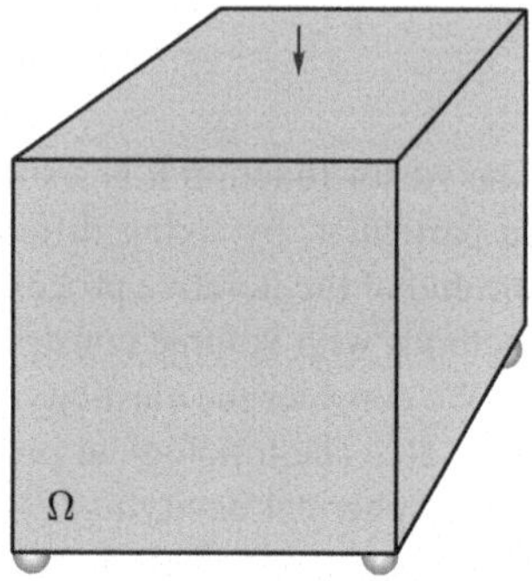

This numerical result is due to the former engineering student Juan Manuel Marmo Lupano and can also be found in [217].

One of the most important requirements in the design of mechanical components is to find an optimal configuration which satisfies a material failure criterion [23, 59, 91, 96, 161, 223]. Following the original ideas presented in [31], let us consider a structural weight minimization problem under stress constraints. We restrict ourselves to the case of the elasticity system in two spatial dimensions, $\mathscr{D} \subset \mathbb{R}^2$. Therefore, given a hold-all domain $\mathscr{D}$ and a stress constraints-enforcement subdomain $\Omega^* \subset \mathscr{D}$, the optimization problem we are dealing with consists in finding a subdomain $\Omega \subset \mathscr{D}$ that solves the following constrained minimization problem:

$$\begin{cases} \underset{\Omega \subset \mathscr{D}}{\text{Minimize}}\ \mathscr{F}_\Omega(u) := |\Omega| + \kappa\, \mathscr{J}(u) \\ \text{subject to} \quad \sigma_M(u) \leq \overline{\sigma} \quad \text{a.e. in} \quad \Omega^* \subset \Omega \end{cases} \tag{6.19}$$

where $\kappa > 0$ and the stress constraints are enforced in the relatively compact subdomain Ω^* of Ω. Finally, u is a solution to the elasticity boundary value problem (6.11) for $d = 2$. Therefore, the idea is to minimize the volume of the structure under local stress constraints. Since this is an ill-posed problem, the shape functional $\mathscr{J}(u)$ in (6.19) represents a regularization term given by the structural compliance, namely

$$\mathscr{J}(u) = \int_{\Gamma_N} \bar{q} \cdot u. \tag{6.20}$$

Some terms in the above expressions still require explanation. The von Mises effective stress $\sigma_M(u)$ is given by

$$\sigma_M(u) := \sqrt{\frac{1}{2}\mathbb{B}\sigma(u) \cdot \sigma(u)} \tag{6.21}$$

with $\mathbb{B} = 3\mathbb{I} - \mathrm{I} \otimes \mathrm{I}$, where $\mathbb{I}$ and I are the fourth and second order identity tensors, respectively. In order to deal with pointwise stress constraints in (6.19), a class of von Mises stress penalty functionals is introduced [31]. It is defined as:

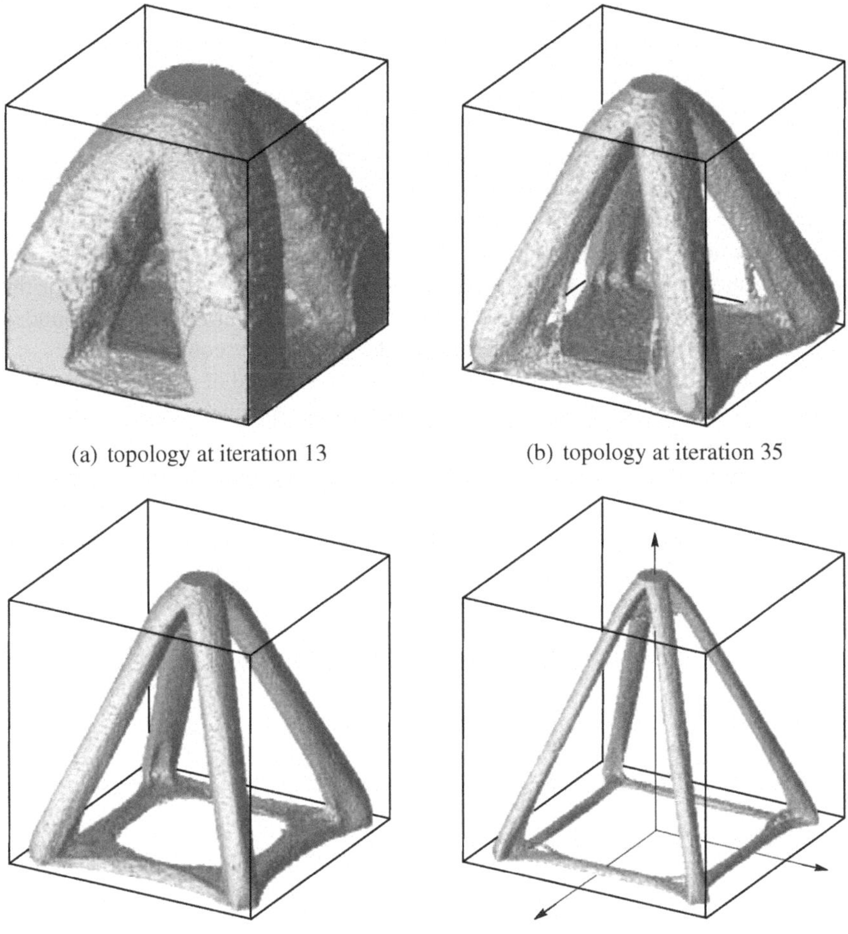

(a) topology at iteration 13

(b) topology at iteration 35

(c) topology at iteration 52

(d) topology at iteration 76

Fig. 6.3 History of the trestle topology design problem [217]

$$\mathcal{G}(u) := \int_{\Omega^*} \Phi_q(\sigma_M^2(u)/\overline{\sigma}^2), \qquad (6.22)$$

where $\Phi_q : \mathbb{R}_+ \to \mathbb{R}_+$ has the following functional form (for more details the reader may refer to [31, 33]):

$$\Phi_q(t) = \left[1 + t^q\right]^{1/q} - 1, \qquad (6.23)$$

where the exponent $q \geq 1$ has to be chosen as large as possible, which is fixed here as $q = 32$. For a detailed explanation of how to choose it we refer to the original paper [23]. Therefore, the previous constrained optimization problem (6.19) can be approximated by the following penalized unconstrained optimization problem:

$$\underset{\Omega \subset \mathscr{D}}{\text{Minimize}} \; \mathscr{F}_{\Omega}^{\alpha}(u) := \mathscr{F}_{\Omega}(u) + \alpha \mathscr{G}(u), \tag{6.24}$$

with the scalar $\alpha > 0$ used to denote a given penalty coefficient. The associated topological derivative $\mathscr{T}(x)$ of (6.24) can be found in [31], for instance. Let us present a numerical example concerning a standard benchmark, which is solved by using Algorithm 1. It consists in the topology design of a structure with a geometrical singularity. The initial guess is given by an L-shaped beam clamped on the top of the vertical branch, which is submitted to a load applied on the top of the horizontal branch, as shown in Fig. 6.4. The final topologies are presented in Fig. 6.5 for the unconstrained and constrained cases. We observe that the reentrant corner is rounded in the constrained case, allowing one to keep the stress under control.

Compliant mechanisms are mechanical devices composed of one single part that transforms simple inputs into complex movements by amplifying and changing their direction [7, 64, 74, 162, 179, 191, 234]. A compliant mechanism needs to be stiff enough to support external loads, and at the same time must be flexible enough to satisfy kinematic requirements. Another difficulty that arises is the tendency of forming flexible joints (hinges), in which the stresses exceed the material failure limit. There are relatively few papers dealing with compliant mechanism design

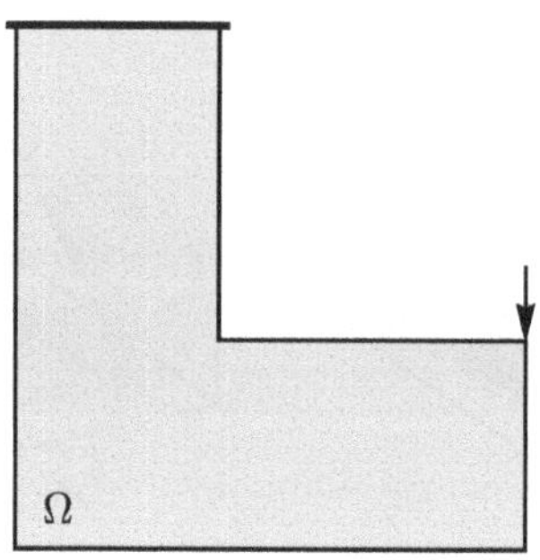

Fig. 6.4 Initial guess and boundary conditions for the L-shaped beam design problem

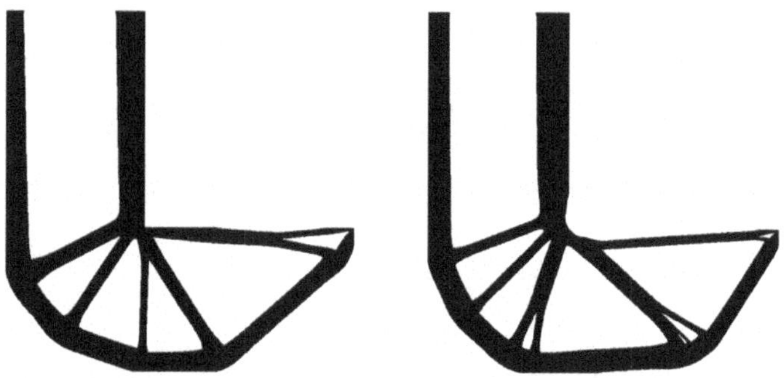

(a) unconstrained (b) constrained

Fig. 6.5 Final configurations for the L-shaped beam design problem [31]

under stress constraints [7, 177, 191]. Following the original ideas from [177], let us consider $\mathscr{D} \subset \mathbb{R}^2$. The Neumann boundary Γ_N consists of three mutually disjoint parts, $\Gamma_N = \Gamma_{in} \cup \Gamma_{out} \cup \Gamma_0$, where input, output and zero boundary tractions are prescribed on Γ_{in}, Γ_{out} and Γ_0, respectively. Therefore, given a hold-all domain $\mathscr{D}$ and a stress constraints-enforcement subdomain $\Omega^* \subset \mathscr{D}$, the optimization problem we are deling with is to find a subdomain $\Omega \subset \mathscr{D}$ that solves the following unconstrained minimization problem:

$$\underset{\Omega \subset \mathscr{D}}{\text{Minimize}}\ \mathscr{F}_\Omega^\alpha(u) := \beta|\Omega| + \mathscr{J}(u) + \alpha\mathscr{G}(u), \tag{6.25}$$

where $\beta > 0, \alpha > 0$ and Ω^* is a relatively compact subdomain of Ω where the stress constraints are enforced. In addition, the von Mises penalty functional $\mathscr{G}(u)$ is given by (6.22). Finally, u is a solution to the elasticity boundary value problem (6.11) for $d = 2$. Since the idea is to maximize the output displacement u_{out} on Γ_{out} for a given input traction on Γ_{in}, the shape functional $\mathscr{J}(u)$ in (6.25) is defined as [25]:

$$\mathscr{J}(u) = \int_{\Gamma_{in}} q_{in} \cdot u + \kappa \int_{\Gamma_{out}} q_{out} \cdot u, \tag{6.26}$$

where q_{in} and q_{out} are given and $\kappa > 0$ is a weight coefficient. The associated topological derivative $\mathscr{T}(x)$ of (6.25) can be found in [177]. To fix ideas, let us present a numerical example, which is solved with the help of Algorithm 1. It consists in an inverter mechanism design. The hold-all domain representing the initial guess is given by a rectangle clamped at the left corners, while the loads q_{in} and q_{out} are respectively applied on the middle of the left and right edges. See Fig. 6.6. The amplified deformations of the final configurations are presented in Fig. 6.7 for the unconstrained and constrained cases. We observe that the resulting mechanisms perform the desired movements. In addition, the constrained case leads to a hinges-free design where the stress is under control.

In most cases of practical interest, the parameters of an optimization problem are not deterministic variables. Applied forces intensities, for example, may not be completely known or may present stochastic variations. Optimization considering

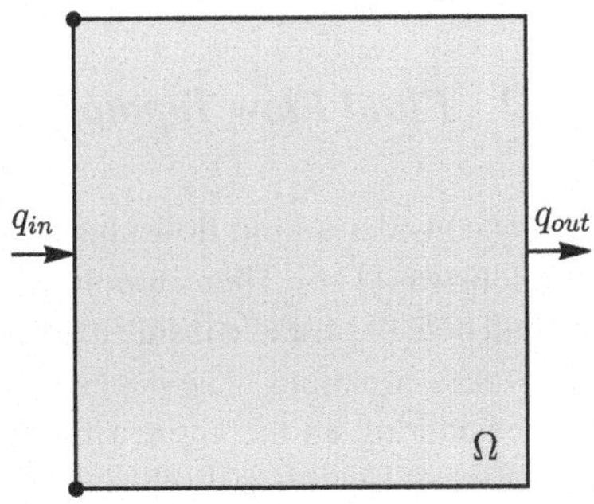

Fig. 6.6 Initial guess and boundary conditions for the inverter design problem

(a) unconstrained

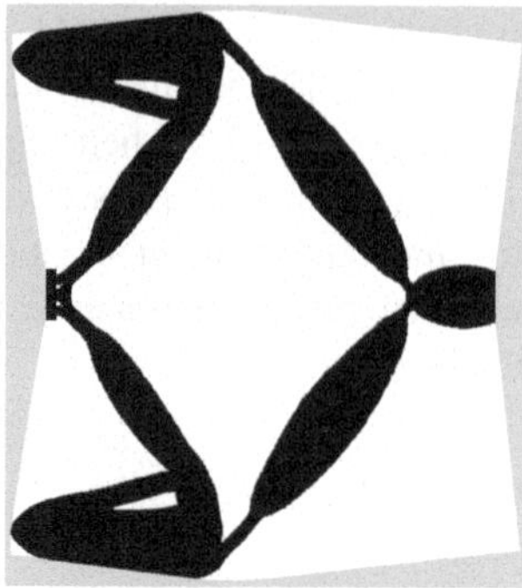

(b) constrained

Fig. 6.7 Amplified deformation of the final configurations for the inverter design problem [177]

uncertainties has been extensively studied in the last decades and several strategies to tackle the problem have been proposed. One interesting branch of research consists in obtaining an optimum design that is least sensitive to variations and uncertainties of the variables, leading to the so-called Robust Optimization. In particular, compliance based robust topology optimization under uncertain loads has been studied in [27, 88, 118, 258], for instance. Here, we follow the ideas presented in [248], where the compliance is evaluated considering a pointwise worst case scenario. Analogously to Sequential Optimization and Reliability Assessment [86], the resulting robust optimization problem can be decoupled into a deterministic topology optimization step and a reliability analysis step. This procedure allows the use of topology optimization algorithms already developed with only small modifications. In particular, since the topological derivative concept has proved to be robust with respect to uncertainties in the data [132], it has been used to address the deterministic topology optimization problem by using Algorithm 1. The reliability analysis step has been handled as in the Performance Measure Approach [251]. Now, let us consider the design of a tower clamped on the bottom and submitted to a pair of uncertain loads, as shown in Fig. 6.8. The optimal topologies considering deterministic and uncertain loading are presented in Fig. 6.9.

6.3.2 *Fluid Flow Topology Design*

Let us consider a fluid flow channel design problem. A first work dealing with this problem was [116]. There, topological sensitivity analysis with respect to insertion of a small hole or obstacle inside a domain has been used to perform shape optimization for Stokes equations. The paper [21] extends this work to Navier–Stokes equations by considering an incompressible fluid and a non-slip condition prescribed on the boundary of an arbitrarily shaped obstacle. So far the methods implemented can only

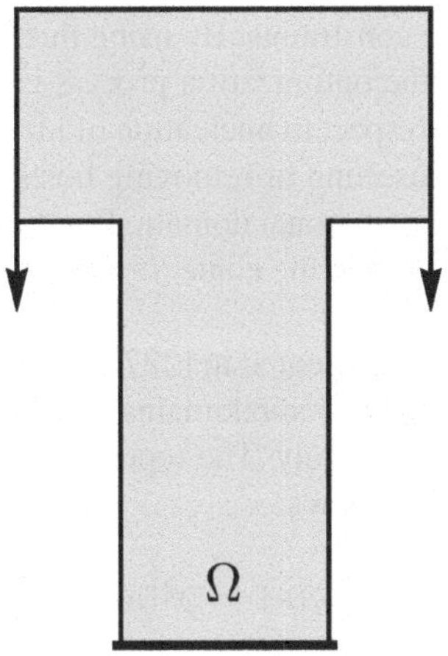

Fig. 6.8 Initial guess and boundary conditions for the tower design problem

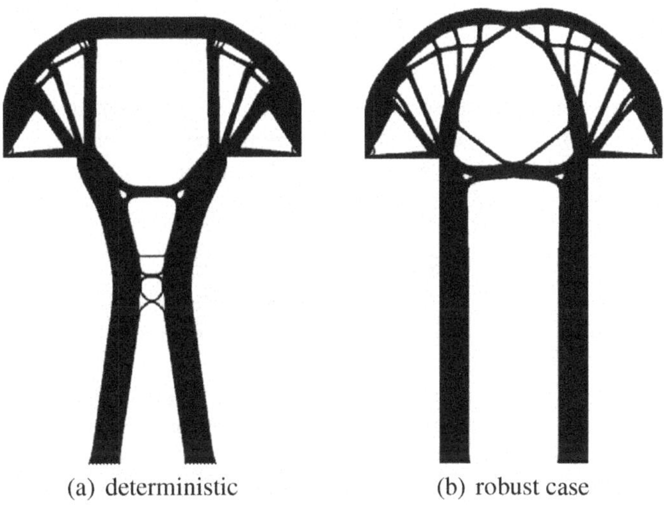

Fig. 6.9 Final configurations for the tower design problem [248]

create small holes inside the domain. Once these holes have been created, they usually remain unchanged during the topological phase of the optimization algorithm. Thus, in [117] a bidirectional topological derivative method is introduced, which allows the decision of whether an existing hole must be removed or not to improve the cost function. In addition, in [87] topological derivative is combined with the standard level-set method for the optimal shape design of Stokes flow. More recently, in [227] a new topological derivative formulation for Stokes as well as Navier–Stokes fluid flow channel design has been proposed, based on traditional topology optimization in which solid or fluid material is distributed at each point of the domain to optimize the cost function subject to some constraints. By using this idea, the problem with hole boundary conditions during the optimization process is solved because asymptotic expansion is performed with respect to nucleation of inclusions – which mimic solid or fluid phases – instead of inserting or removing holes in the fluid domain, which allows working in a fixed computational domain. For the theoretical development of shape and topology optimization in the context of compressible Navier–Stokes see, for instance, [224].

Following the original ideas presented in [227], let us consider a hold-all domain $\mathscr{D} \subset \mathbb{R}^3$, which is divided into two subdomains $\Omega \subset \mathscr{D}$ and $\mathscr{D} \setminus \Omega$, representing the fluid and solid phases, respectively. The topology optimization problem we are dealing with can be written as follows:

$$\underset{\Omega \subset \mathscr{D}}{\text{Minimize}}\ \mathscr{F}_\Omega(u) = \mathscr{J}(u) + \beta|\Omega| \,, \tag{6.27}$$

where $\mathscr{J}(u)$ is the energy shape functional, that is,

$$\mathscr{J}(u) = \mu \int_{\mathscr{D}} \|\nabla u\|^2 + \int_{\mathscr{D}} \alpha \|u\|^2 \,, \tag{6.28}$$

while β is a fixed multiplier used to impose a volume constraint in Ω of the form $|\Omega| \leq M$. Some terms of the energy shape functional require explanation. The function u is a solution of the Navier-Stokes system combined with Darcy's law: Find u and p such that

$$\begin{cases} -\mu\Delta u + (\nabla u)u + \alpha u + \nabla p = 0 & \text{in } \mathscr{D} \,, \\ \operatorname{div}(u) = 0 & \text{in } \mathscr{D} \,, \\ u = u_0 & \text{on } \partial\mathscr{D} \,, \end{cases} \tag{6.29}$$

where $u_0 : \int_{\partial\mathscr{D}} u_0 \cdot \nu = 0$, with ν denoting the outward unit normal to the boundary $\partial\mathscr{D}$. In addition, $0 < \mu < \infty$ is the kinematic viscosity and $\alpha = \alpha(x)$ is the inverse permeability. Therefore, formally $\alpha \to 0$ in Ω and $\alpha \to \infty$ in $\mathscr{D} \setminus \Omega$. This means that α is used to mimic solid or fluid phases, allowing one to work on a fixed computational domain. The topological perturbation we are dealing with is defined by $\alpha_\varepsilon = \gamma_\varepsilon \alpha$, where γ_ε is defined in (6.3). In this case, the topological derivative of the energy shape functional is given by

$$\mathscr{T}_E(x) = -(1-\gamma)\alpha(x)u(x) \cdot (u(x) - v(x)) \quad \forall x \in \mathscr{D} \,, \tag{6.30}$$

while the topological derivative of the volume constraint is trivial. Finally, the auxiliary vector function v is a solution to the following adjoint problem: Find v and q such that

$$\begin{cases} -\mu\Delta v + (\nabla u)^\top v - (\nabla v)u + \alpha v + \nabla q = 2(\alpha u - \mu\Delta u) & \text{in } \mathscr{D} \,, \\ \operatorname{div}(v) = 0 & \text{in } \mathscr{D} \,, \\ v = 0 & \text{on } \partial\mathscr{D} \,. \end{cases} \tag{6.31}$$

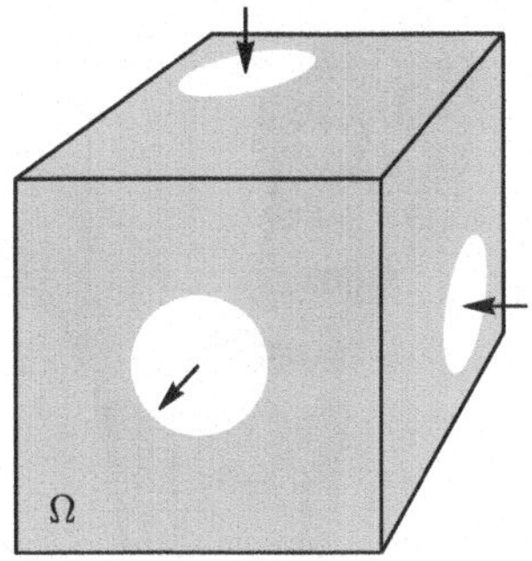

Fig. 6.10 Design domain for the three-way channel design problem

Based on the above ideas, let us present an example of fluid flow design in three spatial dimensions. It is a three-way channel problem, with two inlets, one normal to the X-axis and the other normal to the Y-axis, and one outlet, normal to the Z-axis. Unitary parabolic profiles for the velocity are imposed in the inlets, while in the outlet there is zero pressure. See sketch in Fig. 6.10. The result obtained is shown in Fig. 6.11, where Algorithm 1 has been applied.

6.3.3 Multiscale Material Design

Following the original ideas presented in [109] and further developed in [29], we now present the multiscale material design methodology. It relies on an exact formula for the sensitivity of the macroscopic elasticity tensor to topological microstructural changes. See also [108, 110, 190, 220]. In particular, the associated sensitivity is given by a symmetric fourth order tensor field over the Representative Volume Element (RVE) that measures how the macroscopic elasticity constant estimated within the multiscale framework changes when a small circular hole is introduced at the microscale level. It is derived by making use of the notion of topological derivative within the variational formulation of well-established multiscale constitutive theory, fully developed in the book by Sanchez-Palencia 1980 [231] (see also [106, 193, 194]), where the macroscopic strain and stress tensors are volume averages of their microscopic counterparts over the RVE. The final format of the proposed analytical formula is strikingly simple, so that it is used to devise a topology design algorithm for the synthesis and optimal design of microstructures to meet a specified macroscopic behavior [29]. In particular, we are interested in the synthesis of microstructures of standard material to produce auxetic macrostructures.

Let us consider a macroscopic domain $\Omega \subset \mathbb{R}^2$. Associated with any point $x \in \Omega$ there is a local RVE whose domain is denoted by Ω_μ, with boundary $\partial\Omega_\mu$. The homogenized elasticity tensor $\mathbb{C}$ is defined as

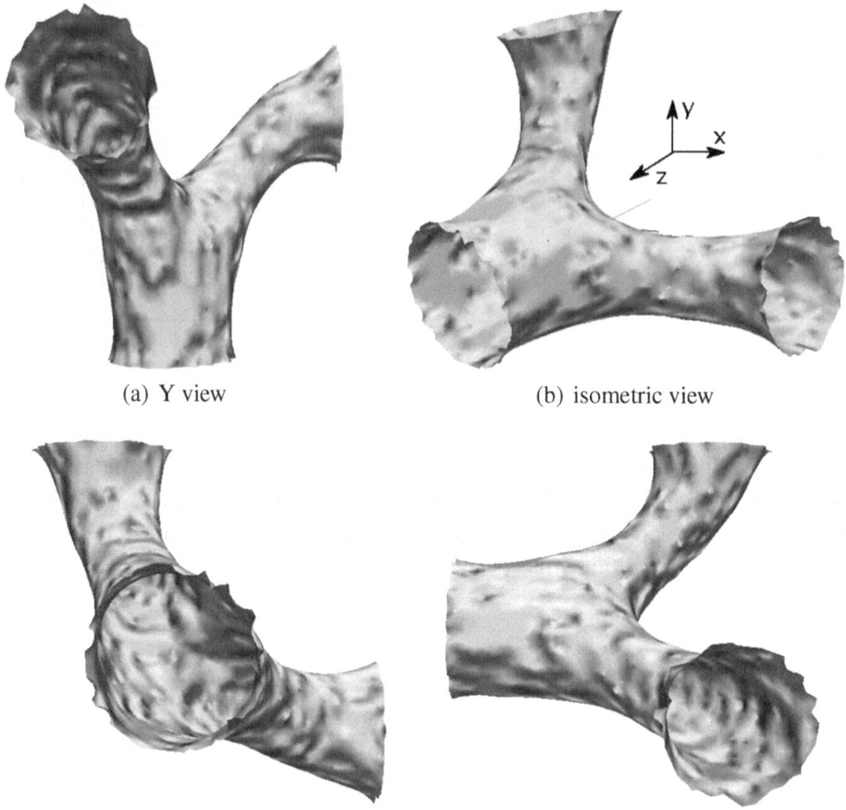

(a) Y view

(b) isometric view

(c) Z view

(d) X view

Fig. 6.11 Final topology (fluid domain) for the three-way channel design problem [227]

$$(\mathbb{C})_{ijkl} = \frac{1}{V_\mu} \int_{\Omega_\mu} (\sigma_\mu(u_\mu^{kl}))_{ij} \;, \tag{6.32}$$

where V_μ denotes the total volume of the RVE and u_μ^{kl} is given by

$$u_\mu^{kl}(y) := u + (e_k \otimes e_l)y + \widetilde{u}_\mu^{kl}(y) \;. \tag{6.33}$$

The constant (rigid) RVE displacement coincides with the macroscopic displacement field u at $x \in \Omega$. The microscopic displacement fluctuation field $\widetilde{u}_\mu^{kl}$ is a solution to the following canonical set of variational problems [231]: Find $\widetilde{u}_\mu^{kl} \in \mathscr{V}_\mu$ such that

$$\int_{\Omega_\mu} \sigma_\mu(\widetilde{u}_\mu^{kl}) \cdot \varepsilon_\mu(\eta) + \int_{\Omega_\mu} \mathbb{C}_\mu(e_k \otimes_s e_l) \cdot \varepsilon_\mu(\eta) = 0 \quad \forall \eta \in \mathscr{V}_\mu \;, \tag{6.34}$$

with $\sigma_\mu(\widetilde{u}^{kl}_\mu) = \mathbb{C}_\mu \varepsilon_\mu(\widetilde{u}^{kl}_\mu)$, where $\varepsilon_\mu(\eta)$ is the symmetric part of the gradient of η. The microscopic constitutive tensor $\mathbb{C}_\mu$ written in terms of the Lamé coefficients μ and λ is given by $\mathbb{C}_\mu = 2\mu\mathbb{I} + \lambda \mathrm{I} \otimes \mathrm{I}$, with I and $\mathbb{I}$ used to denote the second and fourth order identity tensors, respectively. The complete characterization of the multiscale constitutive model is obtained by defining the subspace $\mathscr{V}_\mu \subset \mathscr{U}_\mu$ of kinematically admissible displacement fluctuations. In general, different choices produce different macroscopic responses for the same RVE. In this work we restrict ourselves to media with periodic microstructure. In this case, the geometry of the RVE cannot be arbitrary and must represent a cell whose periodic repetition generates the macroscopic continuum. In addition, displacement fluctuations must satisfy periodicity on the boundary of the RVE. Accordingly, we have

$$\mathscr{V}_\mu := \left\{\varphi \in \mathscr{U}_\mu : \varphi(y^+) = \varphi(y^-) \quad \forall\, (y^+, y^-) \in \mathfrak{P}\right\}, \tag{6.35}$$

where $\mathfrak{P}$ is the set of pairs of points, defined by a one-to-one periodicity correspondence, lying on opposite sides of the RVE boundary. Finally, the minimally constrained space of kinematically admissible displacements $\mathscr{U}_\mu$ is defined as

$$\mathscr{U}_\mu := \left\{\varphi \in H^1(\Omega_\mu) : \int_{\Omega_\mu} \varphi = 0, \ \int_{\partial\Omega_\mu} \varphi \otimes_s n = 0\right\} . \tag{6.36}$$

where n is the outward unit normal to the boundary $\partial\Omega_\mu$ and $\otimes_s$ denotes the symmetric tensor product between vectors. A closed formula for the sensitivity of the homogenized elasticity tensor (6.32) to the nucleation of a circular hole within the RVE has been derived in [109]. It is given by the following fourth order tensor field over Ω_μ:

$$\mathbb{D}_\mu(y) = -\frac{1}{V_\mu}\mathbb{P}_\mu \sigma_\mu(u^{ij}_\mu(y)) \cdot \sigma_\mu(u^{kl}_\mu(y)) e_i \otimes e_j \otimes e_k \otimes e_l \quad \forall y \in \Omega_\mu\,, \tag{6.37}$$

with the polarization tensor $\mathbb{P}_\mu$ given by

$$\mathbb{P}_\mu = \frac{2\mu + \lambda}{3\mu + \lambda}\left(\mathbb{I} + \frac{\mu - \lambda}{4(\mu + \lambda)}\mathrm{I} \otimes \mathrm{I}\right) , \tag{6.38}$$

where the fields u^{ij}_μ come from solutions to (6.34) for the unperturbed RVE domain Ω_μ together with the additive decomposition (6.33). Expression (6.37) allows the exact topological derivative of any differentiable function of $\mathbb{C}$ to be calculated through the direct application of the conventional calculus rules. That is, any such function $\Psi(\mathbb{C})$ has exact topological derivative of the form

$$\mathscr{T}_\mu = \langle D\Psi(\mathbb{C}), \mathbb{D}_\mu \rangle \ , \tag{6.39}$$

with the brackets $\langle\cdot,\cdot\rangle$ denoting the appropriate product between the derivative of Ψ with respect to $\mathbb{C}$ and the topological derivative $\mathbb{D}_\mu$ of $\mathbb{C}$. Note, for example, that properties of interest such as the homogenized Young, shear and bulk moduli as well as the Poisson ratio are all regular functions of $\mathbb{C}$. This fact points strongly to the suitability of the use of (6.39) in a topology design algorithm for the synthesis and optimization of elastic micro-structures based on minimization/maximization of cost functions defined in terms of homogenized properties.

To fix ideas, let us consider a pair $\varphi_1, \varphi_2 \in \mathbb{R}^2 \times \mathbb{R}^2$ of second order tensors. We also introduce the quantity

$$\Psi(\mathbb{C}) := \frac{\mathbb{C}^{-1}\varphi_1 \cdot \varphi_2}{\mathbb{C}^{-1}\varphi_1 \cdot \varphi_1} + \frac{\mathbb{C}^{-1}\varphi_2 \cdot \varphi_1}{\mathbb{C}^{-1}\varphi_2 \cdot \varphi_2} . \tag{6.40}$$

Then the following result, which can be used in numerical methods of synthesis and/or topology design of microstructures [29], holds true:

$$\begin{aligned}\mathcal{T}_\mu = &- \frac{(\mathbb{C}^{-1}\mathbb{D}_\mu\mathbb{C}^{-1})\varphi_1 \cdot [(\mathbb{C}^{-1}\varphi_1 \cdot \varphi_1)\varphi_2 - (\mathbb{C}^{-1}\varphi_1 \cdot \varphi_2)\varphi_1]}{(\mathbb{C}^{-1}\varphi_1 \cdot \varphi_1)^2} \\ &- \frac{(\mathbb{C}^{-1}\mathbb{D}_\mu\mathbb{C}^{-1})\varphi_2 \cdot [(\mathbb{C}^{-1}\varphi_2 \cdot \varphi_2)\varphi_1 - (\mathbb{C}^{-1}\varphi_2 \cdot \varphi_1)\varphi_2]}{(\mathbb{C}^{-1}\varphi_2 \cdot \varphi_2)^2} .\end{aligned} \tag{6.41}$$

Finally, let us consider one RVE given by the unit square $\Omega_\mu = (0, 1) \times (0, 1)$. The initial guess is given by a porous microcell as shown in Fig. 6.12. If we set $\varphi_1 = e_1 \otimes e_1$ and $\varphi_2 = -e_2 \otimes e_2$ in (6.40), the resulting function $\Psi(\mathbb{C})$ yields

$$\Psi(\mathbb{C}) := -\frac{(\mathbb{C}^{-1})_{1122}}{(\mathbb{C}^{-1})_{1111}} - \frac{(\mathbb{C}^{-1})_{1122}}{(\mathbb{C}^{-1})_{2222}} . \tag{6.42}$$

The optimized auxetic microstructure is presented in Fig. 6.13a, while the periodic auxetic structure is shown in Fig. 6.13b. These results have been obtained with the help of Algorithm 1.

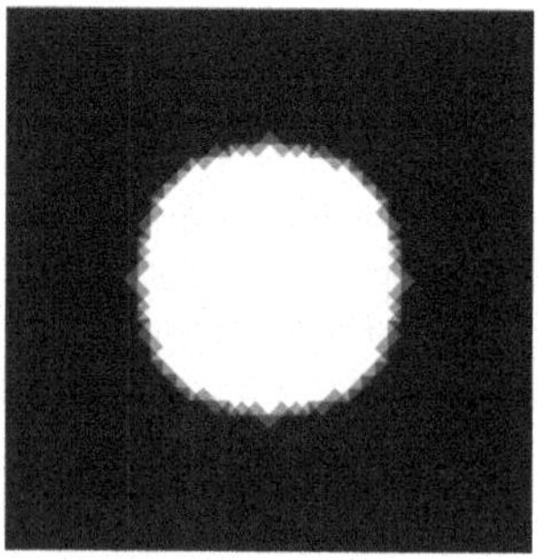

Fig. 6.12 Initial guess for the auxetic microstructure material design

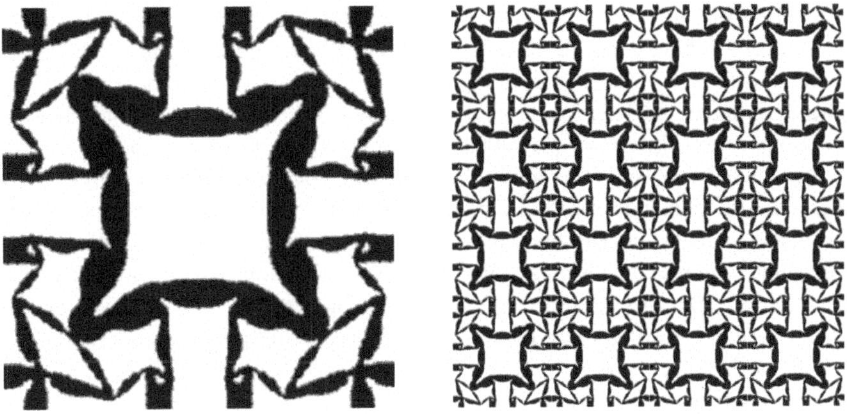

(a) optimized microstructure (b) periodic pattern

Fig. 6.13 Result for the auxetic microstructure material design [29]

6.3.4 Additional Applications

In this section we discuss some additional applications found in the current literature. We do not give details, but the problems are presented in words and precise references are given for the reader convenience.

Inverse imaging problems. Topological derivative has been successfully applied to solve a wide class of inverse problems. The basic idea consists in minimizing a shape functional measuring the misfit between boundary measurements and the solution obtained from the model by using topological derivatives. In particular, the topological derivative field obtained from the background solution gives qualitative information on the shape and topology of hidden defects. Actually, the topological derivative with respect to the nucleation of a small crack embedded into a membrane has been applied in the context of fracture detection from boundary measurements of the associated potential [30]. In addition, topological derivative has also been used to determine the location of small cavities in Stokes flow from velocity boundary measurements [45]. Special attention has been devoted to the topological derivative associated with the Helmholtz problem [230], which has been successfully applied for imaging small acoustic anomalies [28, 75, 97, 104, 119, 141]. See also an experimental validation of the topological derivative method in the context of elastic-wave imaging [247]. The stability and resolution analysis for a topological-derivative-based imaging functional has been presented in [13], showing why it works so well in the context of inverse scattering. See also the related work [120]. For an application of topological derivatives in the context of AFM-based indentation stiffness tomography, see [37]. Finally, applications of topological derivatives in the context of time domain inverse scattering problem can be found in [54, 79, 84], for instance.

Image processing. As a consequence of technological advance, a variety of instruments and tools have been introduced in medicine. For instance, we can refer to medical imaging devices. More specifically, techniques like Computer Tomography, Magnetic Resonance Imaging, Single Photon Emission Tomography, Positron Emission Tomography and Ultrasound provide useful information (anatomical and functional) to specialists, whatever their area of interest (practical medicine, research, etc.). Therefore, the demand for tools to manipulate medical images has grown considerably since the appearance of these technologies. Also different issues have appeared in this field: volume data visualization, image restoration, image segmentation, image registration, pattern recognition and inpainting. The topological derivative concept has been successfully applied to solve image processing problems, including the so-called minimal partition problem [34]. See, for instance, [40, 42, 126, 128, 157]. See also the recent publications [85, 197].

Antenna design in hyperthermia therapy. Hyperthermia is a non-invasive therapy, commonly used in treatment of cancer, consisting in artificially heating body tissues through electromagnetic waves by focusing the heat in cancerous cells. However, one of the challenges in hyperthermia treatment is to selectively heat the cancerous tissue, elevating its temperature above 42 °C, while keeping the temperature of the healthy tissue close to the normal temperature of the human body. In this scenario, the heat applied may damage or even kill the cancerous cells. Even if the cancerous cells do not die immediately, they may become more vulnerable to radiotherapy or chemotherapy, enabling such – in general aggressive – therapies to be given in smaller doses. The regional electromagnetic hyperthermia problem is modeled by a semi-coupled system of partial differential equations. The heat equation in biological tissues, or the *bioheat equation*, is coupled with the *Helmholtz equation*. Electromagnetic waves are generated by a spatially distributed antenna. This antenna produces a source in the Helmholtz equation, whose solution appears as a heat source in the bioheat equation. Therefore, the basic idea consists in finding a distribution of heat source generated by an electromagnetic antenna, which is able to focus the heat into the tumor and keep the temperature under control in the healthy tissue. In other words, the support of the antenna has to be designed, leading to a topology optimization problem. In [9], the problem has been successfully solved with the help of the topological derivative concept. In particular, the authors presented some numerical results showing a possible application of the proposed methodology to cancer treatment by hyperthermia.

Crack nucleation modeling. A simple analytical expression for crack nucleation sensitivity analysis has been derived in [253]. See also [17]. It relies on the concept of topological derivative applied within two-dimensional linear elastic fracture mechanics. In particular, the topological asymptotic expansion of the total potential energy together with a Griffith-type energy of an elastic cracked body has been calculated. As the main result, a crack nucleation criterion based on the topological derivative and a criterion for determining the direction of crack growth based on the associated topological gradient have been introduced.

Damage evolution modeling. The topological derivative associated with the Francfort–Marigo model of damage evolution in brittle materials has been used to nucleate small damaged zones. After nucleating new damages, a level-set method has been used to propagate them according to the associated shape derivative of the Francfort–Marigo functional. For more details, see [8]. However, the whole nucleation and propagation damaging process, including kinking and bifurcation, can also be modeled by using solely the topological derivative concept, leading to a simple and unified approach [256, 257].

6.4 Future Developments

In this chapter the topology optimization algorithm base on the first order topological derivative and the level-set domain representation method has been presented. The large set of applications in the context of topology optimization is provided. General Algorithm 1 has been proposed in [25] to achieve a local optimality condition for the minimization problem under considerations. The local optimality conditions for topology optimization problems are given in terms of the topological derivatives and appropriate level-set functions. This means that the topological derivative is in fact used within the numerical procedure as a steepest-descent direction. Therefore, we propose the class of optimization processes similar to methods based on the gradient of the cost functional. The topological derivative represents the exact first order variation of the shape functional with respect to the nucleation of small singular domain perturbations, so that the resulting topology design algorithm converges in few iterations by using a small number of user defined algorithmic parameters. Furthermore, the topological derivative follows in fact the basic rules of Differential Calculus, which allows for applying it in the context of multi-objective topology optimization algorithms by using e.g., the known formulas already available in the literature. Finally, in contrast to traditional topology optimization methods, the topological derivative formulation does not require any material model concept based on intermediary densities, so that no interpolation schemes are used within the numerical procedures. This feature is crucial in the topology design problem, since the difficulties arising from material model procedures are here naturally avoided. Therefore, the topological derivative method can be seen, when applicable, as a simple alternative method for numerical solution of a wide class of topology optimization problems. For future development of shape-topological first order methods, we highlight the following:

1. According to Sect. 6.3, there is numerical evidence showing that Algorithm 1 converges in most cases. However, from the theoretical point of view, only partial results can be found in the literature. See for instance [24], where the convergence of Algorithm 1 has been analysed in the particular case of an optimal control problem with respect to characteristic functions of small sets. Therefore, the most

important theoretical problem to be solved concerns the convergence analysis of Algorithm 1.

2. Stability and resolution analysis for a topological-derivative-based imaging functional has been presented in the context of the Helmholtz equation [13]. However, such analysis is missing for other classes of inverse problems, including gravimetry and EIT, for instance. In addition, first order topological derivatives are effective in solving a certain class of inverse imaging problems as discussed in Sect. 6.3.4. In Chap. 10 we present a new branch of research, which concerns solving a wide range of inverse reconstruction problems with the help of second order topological derivatives. In this context, many interesting questions arise, including on how to efficiently use higher order expansions, for instance.
3. Synthesis and optimal design of materials in a multiscale framework has been considered in [109], where the topological derivative of the homogenized elasticity tensor has been obtained. Extension to the dynamic case is a difficult and interesting research topic, where inertial forces acting at the microscale may produce unexpected macroscopic constitutive behavior. Finally, a new emerging research field is the design of new materials by considering the strain gradient homogenized constitutive tensor. From the theoretical point of view, a deep question arises in the context of topological derivatives associated with asymptotic models in general, including multiscale and dimension reduction, for instance. In particular, both objects come from a limit passage procedure, one representing the size of the topological perturbation and the other one controlling the scale. It is not clear whether these limits commute or not. Actually, different results should be expected after interchanging the order of these limits.
4. Topology design of structures taking into account more realistic scenario such as anisotropic elasticity, transient wave equations and evolution variational inequalities is a difficult and challenging problem, which requires further development from both theoretical and numerical points of view.
5. Topological-derivative-based topology design in multiphysics taking into account multiobjective shape functionals is an important and difficult subject of research, which also deserves investigation. Design of antenna and wave guides in nanophotonics is an example of modern application. It can be handled with the use of the domain decomposition technique presented in Chap. 2, for instance.
6. The Griffith-Francfort–Marigo damage model adopted in [8] does not distinguish between traction and compression stress states in the damage evolution process. Hence, it is unsuitable to describe the crack closure phenomenon. Therefore, the development of the topological derivative for functionals which specifically would consider distinct criteria in traction and in compression deserves investigation. However, it is well-known that such modeling leads to a class of nonlinear elasticity systems, so that these extensions should be expected to be difficult.

7. Extension to nonlinear problems in general can be considered the main challenge in the theoretical development of the topological derivative method. The difficulty arises when the nonlinearity comes from the main part of the operator, which at the same time suffers a topological perturbation. It is the case of nucleation of holes in plasticity and finite deformations in solid mechanics or small obstacles in compressive fluid flow, for instance. See the recent publication [26] dealing with topological derivatives for a class of quasilinear elliptic equations.

Chapter 7
Synthesis of Compliant Thermomechanical Actuators

This chapter deals with topology design of thermomechanical actuators. The goal of shape optimization is to maximize the output displacement in a given direction on the boundary of the elastic body, which is submitted to a thermal excitation that induces a dilatation/contraction of the thermomechanical device. The optimal structure is identified by an elastic material distribution, while a very compliant (weak) material is used to mimic voids. The mathematical model of an actuator takes the form of a semi-coupled system of partial differential equations. The boundary value problem includes two components, the Navier equation for linear elasticity coupled with the Poisson equation for steady-state heat conduction. The mechanical coupling is the thermal stress induced by the temperature field. Given the integral shape functional, we evaluate its topological derivative with respect to the nucleation of a small circular inclusion with the thermomechanical properties governed by two contrast parameters. The obtained topological derivative is employed to generate a steepest descent direction within the level set numerical procedure of topology optimization in a fixed geometrical domain. Finally, several finite element-based examples for the topology design of thermomechanical actuators are presented. This chapter is based on the paper by Sebastián Miguel Giusti, Zenon Mróz, Antonio André Novotny and Jan Sokolowski [107].

7.1 Preliminaries

In this chapter the topology design of thermomechanical actuators is considered in two spatial dimensions for a linear multiphysics model [149, 174, 226, 234]. The boundary value problem of elliptic type is given by the linearized elasticity coupled with the steady-state heat conduction problem. The reference configuration of the structure is an open and bounded domain $\Omega \subset \mathbb{R}^2$, with Lipschitz boundary denoted as $\partial\Omega$. The topology is identified through the distribution of elastic material within

A. A. Novotny et al., *Applications of the Topological Derivative Method*,
Studies in Systems, Decision and Control 188,
https://doi.org/10.1007/978-3-030-05432-8_7

Ω and the voids are mimicked by a very compliant (weak) material. Therefore, topological changes of the reference domain are defined by the nucleation of inclusions with the thermomechanical properties governed by two contrast parameters. In order to determine the best distribution of elastic material the method of topological derivatives [238] is employed. The shape variations of boundaries and interfaces between elastic and compliant materials are also allowed during the shape-topological optimization. The displacement field in the structure is determined within the framework of linearized elasticity with thermally induced stresses. The temperature field satisfies the steady-state heat conduction equation. The state variables include the displacement field u and the temperature field θ. The shape functional $\Omega \mapsto J(\Omega)$ to be minimized is given by the line integral of $g = -u \cdot e$ on the portion $\Gamma^\star$ of the boundary $\partial\Omega$. Hence, the output displacement u on $\Gamma^\star$ is maximized in a given direction e. For the sake of motivation, let us consider an example which shows that the optimum design of the simple thermomechanical structure does not follow from an intuitive reasoning, namely thermal distortion design of switching device.

7.1.1 Simple Example of a Bar Structure

Consider two bars AP and BP of the length $l = \frac{L}{2}$ and with the joint at P, see Fig. 7.1. The bars are fixed at the points A and B. The length AB of the isosceles triangle APB is $2a$, while a is the design variable. We set $2l = L = const$, where $a = l \cos\beta = \frac{L}{2}\cos\beta$ is varying, and $h = l\sin\beta = \frac{L}{2}\sin\beta$. We want to specify the configuration of two bars for which $\frac{\Delta h}{L}$ is maximum within the admissible configurations of $0 < \beta \le \frac{\pi}{2}$. The structure is uniformly heated up to the temperature θ, which produces a thermal distortion strain given by $\varepsilon_\theta = \alpha\theta$, where α is the thermal expansion coefficient. It is easy to derive the expression for the non-dimensional displacement δ, namely

$$\delta := \frac{\Delta h}{L} = \frac{1}{2}\left[\sqrt{\sin^2\beta + 2\alpha\theta + \alpha^2\theta^2} - \sin\beta\right]$$

and evaluate its derivatives

$$S_\beta := \frac{\partial}{\partial\beta}\left(\frac{\Delta h}{L}\right) = -\frac{\delta\cos\beta}{2\delta + \sin\beta} < 0$$

and

$$S_\theta := \frac{\partial}{\partial\theta}\left(\frac{\Delta h}{L}\right) = \frac{1}{2}\left(\frac{\alpha + \alpha^2\theta}{2\delta + \sin\beta}\right).$$

Since the sensitivity derivative S_β in function of the angle β, decreasing in the interval $(0, \frac{\pi}{2}]$, thus the maximum is located at $\beta = \beta_0$, where $0 < \beta_0 \ll 1$ is a small value of β inducing an upward displacement. In particular, $\beta = \frac{\pi}{2}$ is actually the worst case. It is interesting to note that the maximal value of Δh due to combined rotation and

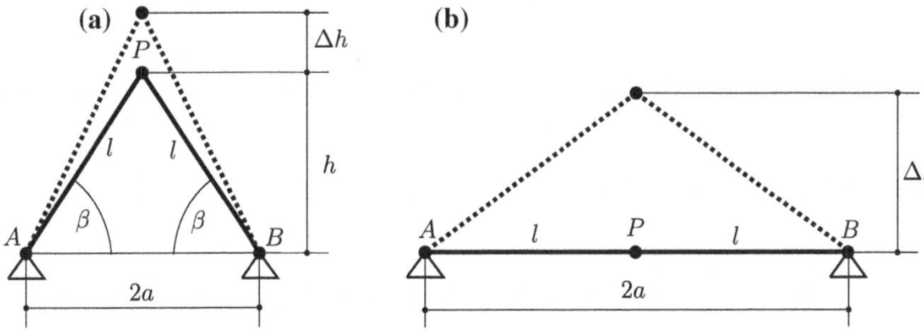

Fig. 7.1 Truss subject to thermal effects: **a** initial guess and **b** optimal layout

extension of bars at $\beta = 0$. This example provides the insight into the principle of optimal design, where the thermal strain induces large rotation of material elements

7.1.2 Topological Derivative for Inclusions

Without loss of generality the topological sensitivity analysis of the given shape functional $\Omega \mapsto J(\Omega)$ can be performed for a single circular inclusion $\varepsilon \mapsto B_\varepsilon(\widehat{x})$. Here $B_\varepsilon(\widehat{x}) = \{\|x - \widehat{x}\| < \varepsilon\}$ is a ball with the fixed centre $\widehat{x} \in \Omega$, so the location of the ball in the reference domain is uniquely determined by its centre. The insertion of inclusion B_ε into the reference domain results in the local perturbation of material properties of the reference domain and it makes the shape functional dependent on the small parameter $\varepsilon \to 0$. In particular, the topologically perturbed counterpart of the shape functional $\varepsilon \mapsto B_\varepsilon \mapsto J_\varepsilon(\Omega)$ is given by the line integral of $g_\varepsilon = -u_\varepsilon \cdot e$ on the portion $\Gamma^\star$ of the boundary $\partial\Omega$. Hence, the output displacement u_ε depends on the inclusion B_ε with the centre $\widehat{x}$ and such that $\varepsilon \to 0$. The dependence of the shape functional results from the state equation, where the small inclusion makes the coefficients of Navier and Poisson equations dependent on the characteristic function of $B_\varepsilon(\widehat{x})$, with $\varepsilon \to 0$ used for the asymptotic analysis. Thus, the interesting question from the point of view of shape-topological optimization is the existence of the asymptotic expansion at $\varepsilon = 0^+$ for the function $\varepsilon \mapsto J_\varepsilon(\Omega)$. Such an asymptotic expansion is established in this chapter. The first term of obtained asymptotic expansion is the so-called topological derivative of the shape functional. The topological derivative depends on the solutions u and θ, as well as on their corresponding adjoint states p and φ, all of them evaluated at the centre $\widehat{x}$. It also depends on the material parameters of the background as well as on γ^M and γ^T, which are called the contrast for mechanical and thermal material properties of B_ε, respectively. In this way an optimal location of a small inclusion and its properties can be determined in order to minimize the shape functional associated with the model. The topological derivative of the elastic energy associated with such a thermomechanical model has been derived by [111]. However, to the best of our knowledge the topological sensitivity analysis of a shape functional specially designed for topology optimization purposes of thermomechanical actuators cannot be found in the literature. Therefore, we derive with all details the topological asymptotic expansion of the adopted shape functional and perform a complete mathematical justification for the obtained formulas.

7.2 Problem Formulation

Let us now introduce the thermomechanical semi-coupled model. The displacement field is determined within linear elasticity with thermally induced stresses for isotropic materials. The temperature field is described by the steady-state heat conduction equation. The state variables include the displacement field and the temperature field.

7.2.1 *Unperturbed Problem*

The shape functional to be minimized is given by a line integral

$$J(\Omega) = \mathcal{J}_\Omega(u) := -\int_{\Gamma^\star} e \cdot u, \tag{7.1}$$

where $\Gamma^\star$ is a part of the boundary $\partial\Omega$ and the structural displacement u has to be maximized in a given unity direction e.

The vector function u solves the following thermomechanical equilibrium problem:

$$u \in \mathcal{V}(\Omega) : \int_\Omega S(u) \cdot (\nabla v)^s = 0 \quad \forall v \in \mathcal{V}(\Omega). \tag{7.2}$$

Some terms in the above equation require explanation. The Cauchy stress tensor $S(u)$ is given by

$$S(u) = \mathbb{C}((\nabla u)^s - \alpha\theta\,\mathrm{I}) = \sigma(u) - \beta\theta\,\mathrm{I}, \tag{7.3}$$

where $(\nabla u)^s$ is used to denote the symmetric part of the gradient of the displacement field u, i.e.

$$(\nabla u)^s := \frac{1}{2}(\nabla u + (\nabla u)^\top). \tag{7.4}$$

In addition, $\mathbb{C}$ denotes the fourth-order elasticity tensor, which for isotropic materials is given by

$$\mathbb{C} = 2\mu\mathbb{I} + \lambda(\mathrm{I} \otimes \mathrm{I}), \tag{7.5}$$

where μ and λ are the Lamé coefficients. The second order tensor $\sigma(u)$ is related to the total displacement field by the Hooke's law

$$\sigma(u) := \mathbb{C}(\nabla u)^s, \tag{7.6}$$

while the coefficient β is given by

$$\beta = \alpha(2\mu + 3\lambda), \tag{7.7}$$

where α is the thermal expansion coefficient. In terms of Young's modulus E and Poisson ratio ν, there are

$$\mu = \frac{E}{2(1+\nu)}, \quad \lambda = \frac{\nu E}{(1+\nu)(1-2\nu)}. \tag{7.8}$$

For plane stress assumption λ and β must be replaced respectively by $\lambda^\star$ in (7.5) and $\beta^\star$ in (7.7), where

$$\lambda^\star = \frac{2\mu\lambda}{\lambda + 2\mu} = \frac{\nu E}{1-\nu^2}, \quad \beta^\star = 2\alpha(\mu + \lambda^\star). \tag{7.9}$$

The space of kinematically admissible displacements is defined as

$$\mathscr{V}(\Omega) := \left\{\phi \in \mathbf{H}^1(\Omega) : \phi_{|\Gamma_u} = 0\right\}, \tag{7.10}$$

with $\mathbf{H}^1(\Omega) := H^1(\Omega; \mathbb{R}^2)$ and Γ_u is used to denote a part of the boundary $\partial\Omega$ where the displacement u is prescribed.

From these elements, the equilibrium equation (7.2) leads to the following variational problem: Find the displacement field $u \in \mathscr{V}(\Omega)$, such that

$$\int_\Omega \sigma(u) \cdot (\nabla v)^s = \int_\Omega \beta\theta \operatorname{div}(v) \quad \forall v \in \mathscr{V}(\Omega), \tag{7.11}$$

where the scalar function θ is the solution to the following variational problem: Find the temperature field $\theta \in \mathscr{H}(\Omega)$, such that

$$\int_\Omega q(\theta) \cdot \nabla\eta + \int_\Omega b\eta = 0 \quad \forall \eta \in \mathscr{H}_0(\Omega), \tag{7.12}$$

with b used to denote a heat source in Ω. The heat flux vector field is defined as

$$q(\theta) = -\mathrm{K}\nabla\theta, \tag{7.13}$$

where K is a second order tensor representing the thermal conductivity of the medium. In the isotropic case, the tensor K can be written as

$$\mathrm{K} = k\mathrm{I}, \tag{7.14}$$

being k the thermal conductivity coefficient. The set $\mathscr{H}(\Omega)$ and the space $\mathscr{H}_0(\Omega)$ are respectively defined as

$$\mathscr{H}(\Omega) := \left\{\phi \in H^1(\Omega) : \phi_{|\Gamma_\theta} = \overline{\theta}\right\}, \tag{7.15}$$

$$\mathscr{H}_0(\Omega) := \left\{\phi \in H^1(\Omega) : \phi_{|\Gamma_\theta} = 0\right\}, \tag{7.16}$$

with Γ_θ used to denote a part of the boundary $\partial\Omega$ where the temperature θ is prescribed by a given function $\overline{\theta}$.

Let us also introduce two adjoint auxiliary problems in order to simplify further analysis. The mechanical auxiliary problem reads: find the adjoint displacement field $p \in \mathscr{V}(\Omega)$, such that

$$\int_\Omega \sigma(p) \cdot (\nabla v)^s = \int_{\Gamma^\star} e \cdot v \quad \forall v \in \mathscr{V}(\Omega). \tag{7.17}$$

The thermal auxiliary problem is stated as: find the adjoint temperature field $\varphi \in \mathscr{H}_0(\Omega)$, such that

$$\int_\Omega q(\varphi) \cdot \nabla\eta = \int_\Omega \beta \operatorname{div}(p)\, \eta \quad \forall \eta \in \mathscr{H}_0(\Omega). \tag{7.18}$$

These adjoint problems result from the Lagrangian formalism, where the associated augmented Lagrangian is minimized with respect to the states u and θ.

7.2.2 Perturbed Problem

The perturbation to the basic problem is now introduced by considering a pair of piecewise constant functions γ_ε^M and γ_ε^T, which respectively affect the constitutive tensors $\mathbb{C}$ and K in some small subdomain of the initial structure. In particular, the topologically perturbed counterpart of the shape functional is given by

$$J_\varepsilon(\Omega) = \mathscr{J}_\Omega(u_\varepsilon) := -\int_{\Gamma^\star} e \cdot u_\varepsilon. \tag{7.19}$$

The vector function u_ε is the solution to the perturbed coupled system, namely: Find the displacement field $u_\varepsilon \in \mathscr{V}(\Omega)$, such that

$$\int_\Omega \sigma_\varepsilon(u_\varepsilon) \cdot (\nabla v)^s = \int_\Omega \beta_\varepsilon \theta_\varepsilon \operatorname{div}(v) \quad \forall v \in \mathscr{V}(\Omega), \tag{7.20}$$

where

$$\sigma_\varepsilon(u_\varepsilon) := \gamma_\varepsilon^M \mathbb{C}(\nabla u_\varepsilon)^s = \gamma_\varepsilon^M \sigma(u_\varepsilon), \tag{7.21}$$

with the contrast on the elastic properties defined as

$$\gamma_\varepsilon^M := \begin{cases} 1 & \text{in } \Omega \setminus \overline{B_\varepsilon} \\ \gamma^M & \text{in } B_\varepsilon \end{cases}. \tag{7.22}$$

Based on the above definition, the perturbed coefficient β_ε in (7.20) takes the form

$$\beta_\varepsilon := \gamma_\varepsilon^M \beta. \tag{7.23}$$

The scalar function θ_ε solves the following perturbed variational problem: Find the temperature field $\theta_\varepsilon \in \mathscr{H}(\Omega)$, such that

$$\int_\Omega q_\varepsilon(\theta_\varepsilon) \cdot \nabla\eta + \int_\Omega b_\varepsilon \eta = 0 \quad \forall \eta \in \mathscr{H}_0(\Omega), \tag{7.24}$$

where

$$q_\varepsilon(\theta_\varepsilon) := -\gamma_\varepsilon^T \mathrm{K}\nabla\theta_\varepsilon, \quad b_\varepsilon := \gamma_\varepsilon^T b, \tag{7.25}$$

with the contrast on the thermal properties defined as

$$\gamma_\varepsilon^T := \begin{cases} 1 & \text{in } \Omega \setminus \overline{B_\varepsilon} \\ \gamma^T & \text{in } B_\varepsilon . \end{cases} \tag{7.26}$$

Finally, the topologically perturbed counterpart of the mechanical adjoint problem (7.17) reads: Find the adjoint displacement field $p_\varepsilon \in \mathscr{V}(\Omega)$, such that

$$\int_\Omega \sigma_\varepsilon(p_\varepsilon) \cdot (\nabla v)^s = \int_{\Gamma^\star} e \cdot v \quad \forall v \in \mathscr{V}(\Omega), \tag{7.27}$$

while the topologically perturbed counterpart of the thermal adjoint problem (7.18) is given by: Find the adjoint temperature field $\varphi_\varepsilon \in \mathscr{H}_0(\Omega)$, such that

$$\int_\Omega q_\varepsilon(\varphi_\varepsilon) \cdot \nabla\eta = \int_\Omega \beta \mathrm{div}(p)\, \eta \quad \forall \eta \in \mathscr{H}_0(\Omega). \tag{7.28}$$

Remark 7.1 Since we are dealing with a topology optimization problem in a fixed domain Ω, the optimal structure is identified by the elastic material, while a very compliant (weak) material is used to mimic voids, both distributed within Ω. Therefore, we consider contrasts in the Young's modulus E (assuming the Poisson ratio ν as constant) and in the thermal conductivity k only, which are respectively given by γ^M and γ^T. In fact, by setting $\gamma^M \to 0$ and $\gamma^T \to 0$, the transmission conditions on $\partial B_\varepsilon(\widehat{x})$ degenerate to homogeneous Neumann boundary conditions in both mechanical (7.20) and thermal (7.24) problems, respectively representing a void and an ideal thermal insulation. The general case is much more involved, so that we left it for future work.

7.3 Existence of the Topological Derivative

The following results ensure the existence of the topological derivative associated with the problem under analysis.

Lemma 7.1 *Let θ and θ_ε be solutions to* (7.12) *and* (7.24), *respectively. Then we have that the following estimate holds true*

$$\|\theta_\varepsilon - \theta\|_{H^1(\Omega)} \leq C\varepsilon. \tag{7.29}$$

Proof We start by subtracting the variational problem (7.12) from (7.24). After some manipulations there is:

$$\int_\Omega q_\varepsilon(\theta_\varepsilon - \theta) \cdot \nabla\eta = (1-\gamma^T)\int_{B_\varepsilon} q(\theta)\cdot\nabla\eta + (1-\gamma^T)\int_{B_\varepsilon} b\eta, \tag{7.30}$$

where we have used the fact that $q_\varepsilon(\phi) = q(\phi)$ and $b_\varepsilon = b$ in $\Omega \setminus \overline{B_\varepsilon}$, and $q_\varepsilon(\phi) = \gamma^T q(\phi)$ and $b_\varepsilon = \gamma^T b$ in B_ε. By taking $\eta = \theta_\varepsilon - \theta$ as a test function in the above equation we obtain the following equality

$$\int_\Omega q_\varepsilon(\theta_\varepsilon - \theta)\cdot\nabla(\theta_\varepsilon - \theta) = (1-\gamma^T)\int_{B_\varepsilon} q(\theta)\cdot\nabla(\theta_\varepsilon-\theta) + \\ (1-\gamma^T)\int_{B_\varepsilon} b(\theta_\varepsilon - \theta). \tag{7.31}$$

From the Cauchy–Schwartz inequality it follows that

$$\int_\Omega q_\varepsilon(\theta_\varepsilon - \theta)\cdot\nabla(\theta_\varepsilon-\theta) \leq C_1\|q(\theta)\|_{L^2(B_\varepsilon)}\|\nabla(\theta_\varepsilon-\theta)\|_{L^2(B_\varepsilon)} + \\ C_2\|b\|_{L^2(B_\varepsilon)}\|\theta_\varepsilon - \theta\|_{L^2(B_\varepsilon)} \leq \varepsilon C_3\|\theta_\varepsilon - \theta\|_{H^1(\Omega)}, \tag{7.32}$$

where we have used the interior elliptic regularity of function θ and the continuity of the function b at the point $\widehat{x} \in \Omega$. Finally, from the coercivity of the bilinear form on the left-hand side of (7.30), namely

$$c\|\theta_\varepsilon - \theta\|^2_{H^1(\Omega)} \leq \int_\Omega q_\varepsilon(\theta_\varepsilon-\theta)\cdot\nabla(\theta_\varepsilon-\theta), \tag{7.33}$$

we obtain the result with the constant $C = C_3/c$ independent of the small parameter ε.

Lemma 7.2 *Let u and u_ε be solutions to* (7.11) *and* (7.20), *respectively. Then we have that the following estimate holds true*

$$\|u_\varepsilon - u\|_{\mathbf{H}^1(\Omega)} \leq C\varepsilon. \tag{7.34}$$

Proof Let us subtract the variational problem (7.11) from (7.20), so that after some manipulations we have:

$$\begin{aligned}\int_\Omega \sigma_\varepsilon(u_\varepsilon - u) \cdot (\nabla v)^s = \int_\Omega \beta(\theta_\varepsilon - \theta)\mathrm{div}(v) + \\ (1-\gamma^M)\int_{B_\varepsilon} (\sigma(u) + \beta\theta\,\mathrm{I}) \cdot (\nabla v)^s - \\ (1-\gamma^M)\int_{B_\varepsilon} \beta(\theta_\varepsilon - \theta)\mathrm{div}(v),\end{aligned} \tag{7.35}$$

where we have used the fact that $\sigma_\varepsilon(\phi) = \sigma(\phi)$ and $\beta_\varepsilon = \beta$ in $\Omega \setminus \overline{B_\varepsilon}$, and $\sigma_\varepsilon(\phi) = \gamma^M\sigma(\phi)$ and $\beta_\varepsilon = \gamma^M\beta$ in B_ε. By taking $v = u_\varepsilon - u$ as test function in the above equation we obtain the following equality

$$\begin{aligned}\int_\Omega \sigma_\varepsilon(u_\varepsilon - u) \cdot (\nabla(u_\varepsilon - u))^s = \int_\Omega \beta(\theta_\varepsilon - \theta)\mathrm{div}(u_\varepsilon - u) + \\ (1-\gamma^M)\int_{B_\varepsilon} (\sigma(u) + \beta\theta\,\mathrm{I}) \cdot (\nabla(u_\varepsilon - u))^s - \\ (1-\gamma^M)\int_{B_\varepsilon} \beta(\theta_\varepsilon - \theta)\mathrm{div}(u_\varepsilon - u).\end{aligned} \tag{7.36}$$

From the Cauchy–Schwartz inequality it follows that

$$\begin{aligned}\int_\Omega \sigma_\varepsilon(u_\varepsilon - u) \cdot (\nabla(u_\varepsilon - u))^s \leq C_1\|\theta_\varepsilon - \theta\|_{L^2(\Omega)}\|\nabla(u_\varepsilon - u)\|_{\mathbf{L}^2(\Omega)} + \\ C_2\|\sigma(u) + \beta\theta\,\mathrm{I}\|_{\mathbf{L}^2(B_\varepsilon)}\|\nabla(u_\varepsilon - u)\|_{\mathbf{L}^2(B_\varepsilon)} + C_3\|\theta_\varepsilon - \theta\|_{L^2(B_\varepsilon)}\|\nabla(u_\varepsilon - u)\|_{\mathbf{L}^2(B_\varepsilon)} \leq \\ C_4\|\theta_\varepsilon - \theta\|_{H^1(\Omega)}\|u_\varepsilon - u\|_{\mathbf{H}^1(\Omega)} + \varepsilon C_5\|u_\varepsilon - u\|_{\mathbf{H}^1(\Omega)},\end{aligned} \tag{7.37}$$

where we have used the interior elliptic regularity of function u and the continuity of the function β at the point $\widehat{x} \in \Omega$. From Lemma 7.1 we have now

$$\int_\Omega \sigma_\varepsilon(u_\varepsilon - u) \cdot (\nabla(u_\varepsilon - u))^s \leq C_6\varepsilon\|u_\varepsilon - u\|_{\mathbf{H}^1(\Omega)}. \tag{7.38}$$

Finally, from the coercivity of the bilinear form on the left-hand side of (7.35), namely

$$c\|u_\varepsilon - u\|^2_{\mathbf{H}^1(\Omega)} \leq \int_\Omega \sigma_\varepsilon(u_\varepsilon - u) \cdot (\nabla(u_\varepsilon - u))^s, \tag{7.39}$$

we obtain the result with the constant $C = C_6/c$ independent of the small parameter ε.

Lemma 7.3 *Let p and p_ε be solutions to* (7.17) *and* (7.27), *respectively. Then we have that the following estimate holds true*

$$\|p_\varepsilon - p\|_{\mathbf{H}^1(\Omega)} \leq C\varepsilon. \tag{7.40}$$

Proof After subtracting the variational problem (7.17) from (7.27) we have:

$$\int_\Omega \sigma_\varepsilon(p_\varepsilon - p) \cdot (\nabla v)^s = (1 - \gamma^M) \int_{B_\varepsilon} \sigma(p) \cdot (\nabla v)^s, \tag{7.41}$$

where we have used the fact that $\sigma_\varepsilon(\phi) = \sigma(\phi)$ in $\Omega \setminus \overline{B_\varepsilon}$ and $\sigma_\varepsilon(\phi) = \gamma^M \sigma(\phi)$ in B_ε. By taking $v = p_\varepsilon - p$ as test function in the above equation we obtain the following equality

$$\int_\Omega \sigma_\varepsilon(p_\varepsilon - p) \cdot (\nabla(p_\varepsilon - p))^s = (1 - \gamma^M) \int_{B_\varepsilon} \sigma(p) \cdot (\nabla(p_\varepsilon - p))^s. \tag{7.42}$$

From the Cauchy–Schwartz inequality it follows that

$$\int_\Omega \sigma_\varepsilon(p_\varepsilon - p) \cdot (\nabla(p_\varepsilon - p))^s \leq$$
$$C_1 \|\sigma(p)\|_{\mathbf{L}^2(B_\varepsilon)} \|\nabla(p_\varepsilon - p)\|_{\mathbf{L}^2(B_\varepsilon)} \leq \varepsilon C_2 \|p_\varepsilon - p\|_{\mathbf{H}^1(\Omega)}, \tag{7.43}$$

where we have used the interior elliptic regularity of function p. Finally, from the coercivity of the bilinear form on the left-hand side of (7.41), namely

$$c\|p_\varepsilon - p\|^2_{\mathbf{H}^1(\Omega)} \leq \int_\Omega \sigma_\varepsilon(p_\varepsilon - p) \cdot (\nabla(p_\varepsilon - p))^s, \tag{7.44}$$

we obtain the result with the constant $C = C_2/c$ independent of the small parameter ε.

Lemma 7.4 *Let φ and φ_ε be solutions to* (7.18) *and* (7.28), *respectively. Then we have that the following estimate holds true*

$$\|\varphi_\varepsilon - \varphi\|_{H^1(\Omega)} \leq C\varepsilon. \tag{7.45}$$

Proof After subtracting the variational problem (7.18) from (7.28) there is:

$$\int_{\Omega} q_\varepsilon(\varphi_\varepsilon - \varphi) \cdot \nabla \eta = (1 - \gamma^T) \int_{B_\varepsilon} q(\varphi) \cdot \nabla \eta, \tag{7.46}$$

where we have used the fact that $q_\varepsilon(\phi) = q(\phi)$ in $\Omega \setminus \overline{B_\varepsilon}$ and $q_\varepsilon(\phi) = \gamma^T q(\phi)$ in B_ε. By taking $\eta = \varphi_\varepsilon - \varphi$ as test function in the above equation we obtain the following equality

$$\int_{\Omega} q_\varepsilon(\varphi_\varepsilon - \varphi) \cdot \nabla(\varphi_\varepsilon - \varphi) = (1 - \gamma^T) \int_{B_\varepsilon} q(\varphi) \cdot \nabla(\varphi_\varepsilon - \varphi). \tag{7.47}$$

From the Cauchy–Schwartz inequality it follows that

$$\int_{\Omega} q_\varepsilon(\varphi_\varepsilon - \varphi) \cdot \nabla(\varphi_\varepsilon - \varphi) \leq$$
$$C_1 \|q(\varphi)\|_{L^2(B_\varepsilon)} \|\nabla(\varphi_\varepsilon - \varphi)\|_{L^2(B_\varepsilon)} \leq \varepsilon C_2 \|\varphi_\varepsilon - \varphi\|_{H^1(\Omega)}, \tag{7.48}$$

where we have used the interior elliptic regularity of function φ. Finally, from the coercivity of the bilinear form on the left-hand side of (7.46), namely

$$c\|\varphi_\varepsilon - \varphi\|^2_{H^1(\Omega)} \leq \int_{\Omega} q_\varepsilon(\varphi_\varepsilon - \varphi) \cdot \nabla(\varphi_\varepsilon - \varphi), \tag{7.49}$$

we obtain the result with the constant $C = C_2/c$ independent of the small parameter ε.

7.4 Topological Asymptotic Analysis

Since the problem under consideration is linear, we firstly set $\gamma^M = 1$ and develop the topological asymptotic analysis for $\gamma^T \neq 1$. Next, we set $\gamma^T = 1$ and develop the analysis for $\gamma^M \neq 1$. Finally, the obtained results are superposed, leading to the associated topological derivative for any pair of γ^M and γ^T. Let us start by evaluating the difference between the original and perturbed shape functionals given respectively by (7.1) and (7.19), which leads to

$$\mathcal{J}_\Omega(u_\varepsilon) - \mathcal{J}_\Omega(u) = -\int_{\Gamma^\star} e \cdot (u_\varepsilon - u). \tag{7.50}$$

7.4.1 *Contrast on the Elastic Coefficients*

Let us set $\gamma^T = 1$ and develop the analysis for $\gamma^M \neq 1$. In this case we have immediately that $\theta_\varepsilon \equiv \theta$. By taking $v = u_\varepsilon - u$ in (7.27), we obtain the equality

$$\int_{\Omega} \sigma_{\varepsilon}(p_{\varepsilon}) \cdot (\nabla(u_{\varepsilon} - u))^{s} = \int_{\Gamma^{\star}} e \cdot (u_{\varepsilon} - u). \tag{7.51}$$

Now, let us set $v = p_{\varepsilon}$ in (7.11) and (7.20). After evaluating the difference between the obtained results we get

$$\int_{\Omega} \sigma_{\varepsilon}(p_{\varepsilon}) \cdot (\nabla(u_{\varepsilon} - u))^{s} = \frac{1-\gamma^{M}}{\gamma^{M}} \int_{B_{\varepsilon}} \sigma_{\varepsilon}(p_{\varepsilon}) \cdot (\nabla u)^{s} - (1-\gamma^{M}) \int_{B_{\varepsilon}} \beta\theta \operatorname{div}(p_{\varepsilon}). \tag{7.52}$$

Therefore, after comparing the last two results with (7.50) we have

$$\mathcal{J}_{\Omega}(u_{\varepsilon}) - \mathcal{J}_{\Omega}(u) = -\frac{1-\gamma^{M}}{\gamma^{M}} \int_{B_{\varepsilon}} \sigma_{\varepsilon}(p_{\varepsilon}) \cdot (\nabla u)^{s} + (1-\gamma^{M}) \int_{B_{\varepsilon}} \beta\theta \operatorname{div}(p_{\varepsilon}). \tag{7.53}$$

Let us propose an ansätz for p_{ε} in the form $p_{\varepsilon} = p + w_{\varepsilon} + \widetilde{p}_{\varepsilon}$, which allows us to choose w_{ε} as a solution to: Find the exterior displacement field $w_{\varepsilon} \in \mathbf{W}^{1}(\mathbb{R}^{2})/\mathbb{R}$, such that

$$\int_{\mathbb{R}^{2}} \sigma_{\varepsilon}(w_{\varepsilon}) \cdot (\nabla v)^{s} = (1-\gamma^{M})\sigma(p)(\widehat{x}) \cdot \int_{B_{\varepsilon}} (\nabla v)^{s} \quad \forall v \in \mathbf{W}^{1}(\mathbb{R}^{2})/\mathbb{R}. \tag{7.54}$$

where the weighted quotient space $\mathbf{W}^{1}(\mathbb{R}^{2})/\mathbb{R}$ has been introduced in [35, App. C] to ensure existence and uniqueness of a solution to the above exterior problem. From Eshelby's Theorem [94, 95], the exterior problem (7.54) admits an explicit solution, namely

$$\sigma_{\varepsilon}(w_{\varepsilon}) = \mathbb{T}\sigma(p)(\widehat{x}) \quad \text{in} \quad B_{\varepsilon}, \tag{7.55}$$

where $\mathbb{T}$ is a fourth order isotropic tensor written as

$$\mathbb{T} = \gamma^{M} \frac{1-\gamma^{M}}{1+\gamma^{M}\alpha_{2}} \left(\alpha_{2}\mathbb{I} + \frac{\alpha_{1}-\alpha_{2}}{2(1+\gamma^{M}\alpha_{1})} \mathrm{I} \otimes \mathrm{I} \right), \tag{7.56}$$

with the constants α_1 and α_2 given by

$$\alpha_{1} = \frac{\mu+\lambda}{\mu}, \quad \alpha_{2} = \frac{3\mu+\lambda}{\mu+\lambda}. \tag{7.57}$$

See, for instance, the book [219, Ch. 5, pp. 156]. From Lemma 7.3, the remainder $\widetilde{p}_{\varepsilon}$ has an estimate of the form $\|\widetilde{p}_{\varepsilon}\|_{\mathbf{H}^{1}(\Omega)} \approx o(\varepsilon)$. Finally, by taking into account these last results, we have the following expansion for the shape functional

$$\mathscr{J}_{\Omega}(u_\varepsilon) - \mathscr{J}_{\Omega}(u) = -\pi\varepsilon^2 \mathbb{P}\sigma(u)(\widehat{x}) \cdot (\nabla p)^s(\widehat{x}) + \\ \pi\varepsilon^2\beta(1+\alpha_1)\frac{1-\gamma^M}{1+\gamma^M\alpha_1}\theta(\widehat{x})\mathrm{div}(p)(\widehat{x}) + o(\varepsilon^2). \quad (7.58)$$

where $\mathbb{P}$ is the Polya–Szëgo polarization tensor given by [16]

$$\mathbb{P} = \frac{1-\gamma^M}{1+\gamma^M\alpha_2}\left((1+\alpha_2)\mathbb{I} + \frac{1}{2}(\alpha_1-\alpha_2)\frac{1-\gamma^M}{1+\gamma^M\alpha_1}\mathrm{I}\otimes\mathrm{I}\right). \quad (7.59)$$

7.4.2 Contrast on the Thermal Coefficients

Let us now set $\gamma^M = 1$ and develop the analysis for $\gamma^T \neq 1$. By taking $v = u_\varepsilon - u$ in (7.27), we obtain

$$\int_\Omega \sigma(p) \cdot (\nabla(u_\varepsilon - u))^s = \int_{\Gamma^\star} e \cdot (u_\varepsilon - u). \quad (7.60)$$

Now, let us set $v = p$ in (7.11) and (7.20). After evaluating the difference between the obtained results we get

$$\int_\Omega \sigma(u_\varepsilon - u) \cdot (\nabla p)^s = \int_\Omega \beta(\theta_\varepsilon - \theta)\mathrm{div}(p). \quad (7.61)$$

Therefore, after comparing the last two results with (7.50) we have

$$\mathscr{J}_{\Omega}(u_\varepsilon) - \mathscr{J}_{\Omega}(u) = -\int_\Omega \beta(\theta_\varepsilon - \theta)\mathrm{div}(p). \quad (7.62)$$

By setting $\eta = \theta_\varepsilon - \theta$ in (7.28) we obtain the following equality

$$\int_\Omega q_\varepsilon(\varphi_\varepsilon) \cdot \nabla(\theta_\varepsilon - \theta) = \int_\Omega \beta(\theta_\varepsilon - \theta)\mathrm{div}(p), \quad (7.63)$$

which leads to

$$\mathscr{J}_{\Omega}(u_\varepsilon) - \mathscr{J}_{\Omega}(u) = -\int_\Omega q_\varepsilon(\varphi_\varepsilon) \cdot \nabla(\theta_\varepsilon - \theta). \quad (7.64)$$

Now, let us set $\eta = \varphi_\varepsilon$ in (7.12) and (7.24). After evaluating the difference between the obtained results we get

$$\int_\Omega q_\varepsilon(\theta_\varepsilon - \theta) \cdot \nabla\varphi_\varepsilon = (1-\gamma^T)\int_{B_\varepsilon} q(\theta) \cdot \nabla\varphi_\varepsilon + (1-\gamma^T)\int_{B_\varepsilon} b\varphi_\varepsilon. \quad (7.65)$$

After comparing the last two results with (7.64) we finally obtain

$$\mathcal{J}_\Omega(u_\varepsilon) - \mathcal{J}_\Omega(u) = -\frac{1-\gamma^T}{\gamma^T}\int_{B_\varepsilon} q_\varepsilon(\varphi_\varepsilon)\cdot\nabla\theta - (1-\gamma^T)\int_{B_\varepsilon} b\varphi_\varepsilon. \quad (7.66)$$

Let us propose an ansätz for φ_ε in the form $\varphi_\varepsilon = \varphi + \vartheta_\varepsilon + \widetilde{\varphi}_\varepsilon$, which allows us to choose ϑ_ε as a solution to: Find the exterior temperature field $\vartheta_\varepsilon \in W^1(\mathbb{R}^2)/\mathbb{R}$, such that

$$\int_{\mathbb{R}^2} q_\varepsilon(\vartheta_\varepsilon)\cdot\nabla\eta = (1-\gamma^T)q(\theta)(\widehat{x})\cdot\int_{B_\varepsilon}\nabla\eta \quad \forall\eta\in W^1(\mathbb{R}^2)/\mathbb{R}, \quad (7.67)$$

where the weighted quotient space $W^1(\mathbb{R}^2)/\mathbb{R}$ has been introduced in [35, App. C] to ensure existence and uniqueness of a solution to the above exterior problem. The exterior problem (7.67) admits an explicit solution, namely

$$q_\varepsilon(\vartheta_\varepsilon) = \gamma^T\frac{1-\gamma^T}{1+\gamma^T}q(\varphi)(\widehat{x}) \quad \text{in} \quad B_\varepsilon. \quad (7.68)$$

See, for instance, the book [219, Ch. 5, pp. 144]. From Lemma 7.4, we have that the remainder $\widetilde{\varphi}_\varepsilon$ has an estimate of the form $\|\widetilde{\varphi}_\varepsilon\|_{H^1(\Omega)} \approx o(\varepsilon)$. From these last results, we have the following expansion for the shape functional

$$\mathcal{J}_\Omega(u_\varepsilon) - \mathcal{J}_\Omega(u) = -\pi\varepsilon^2 \mathrm{P}q(\theta)(\widehat{x})\cdot\nabla\varphi(\widehat{x}) - \pi\varepsilon^2(1-\gamma^T)b\varphi(\widehat{x}) + o(\varepsilon^2), \quad (7.69)$$

where P is the Polya–Szëgo polarization tensor given by [16]

$$\mathrm{P} = 2\frac{1-\gamma^T}{1+\gamma^T}\mathrm{I}. \quad (7.70)$$

7.4.3 *Topological Derivative*

Since the problem is linear, we can sum the obtained expansions (7.58) and (7.69) to obtain

$$\begin{aligned}\mathcal{J}_\Omega(u_\varepsilon) - \mathcal{J}_\Omega(u) = {}& -\pi\varepsilon^2\mathbb{P}\sigma(u)(\widehat{x})\cdot(\nabla p)^s(\widehat{x}) + \\ & \pi\varepsilon^2\beta(1+\alpha_1)\frac{1-\gamma^M}{1+\gamma^M\alpha_1}\theta(\widehat{x})\mathrm{div}(p)(\widehat{x}) - \\ & \pi\varepsilon^2\mathrm{P}q(\theta)(\widehat{x})\cdot\nabla\varphi(\widehat{x}) - \pi\varepsilon^2(1-\gamma^T)b\varphi(\widehat{x}) + o(\varepsilon^2), \quad (7.71)\end{aligned}$$

which ensures the existence for the topological derivative of the shape functional $J(\Omega)$ for $f(\varepsilon) = \pi\varepsilon^2$, provided that the remainder has order $o(\varepsilon^2)$. Finally, the topological derivative can be promptly identified, which is given by the following closed formula

$$\mathscr{T}(\widehat{x}) = -\mathbb{P}\sigma(u)(\widehat{x}) \cdot (\nabla p)^s(\widehat{x}) + \beta(1+\alpha_1)\frac{1-\gamma^M}{1+\gamma^M\alpha_1}\theta(\widehat{x})\mathrm{div}(p)(\widehat{x}) - \mathrm{P}q(\theta)(\widehat{x}) \cdot \nabla\varphi(\widehat{x}) - (1-\gamma^T)b\varphi(\widehat{x}). \tag{7.72}$$

Notice that the closed formula for the topological derivative depends only on the solution of the direct and adjoint problems, given by Eqs. (7.11), (7.12), (7.17) and (7.18), evaluated at the point $\widehat{x}$. This derivative represents the sensitivity of the multi-physics problem, we are dealing with, to the insertion of a circular inclusion of radius ε and center at an arbitrary point $\widehat{x} \in \Omega$, whose constitutive properties are characterized by the contrasts γ^M and γ^T.

7.5 Numerical Experiments

In this section two numerical examples of topology design of thermomechanical actuators into plane stress assumptions are presented. The topology design algorithm developed by [25] is adopted in order to solve the optimization problem, which is based on the topological derivative concept together with a level-set domain representation method. For further details of the algorithm we refer to the work by [31]. See also Algorithm 1 presented in Chap. 6. In all examples we consider the following constitutive properties: $E = 1$ GPa (Young's modulus), $\nu = 0.3$ (Poisson's ratio), $\alpha = 1.0 \times 10^{-6}\,\mathrm{K}^{-1}$ and $k = 1.0\,\mathrm{W/mK}$. The contrast parameters are given by $\gamma^M = \gamma^T = 1.0 \times 10^{-4}$, which are used to mimic the voids. In the part of the boundary where nothing is specified, we consider homogeneous Neumann boundary conditions in both problems (mechanical and thermal). The direction e is given by a unit vector on $\Gamma^\star$. In addition, we do not consider a heat source, i.e. $b = 0$.

The thermomechanical problem (7.11), the steady-state heat conduction problem (7.12) and the adjoint equations (7.17) and (7.18) are solved by using the standard finite element method [259]. The initial mesh is generated from a regular grid of size 20×12, where each resulting square is divided into four triangles, leading to 960 elements. Then, four steps of uniform mesh refinement are performed during the iterative process. In the figures, black and white are respectively used to represent solid and void, whereas the color levels black/brown to yellow/white indicate colder to hotter, respectively. Finally, the procedures described by [66] were used to impose a targeted final volume.

In order to analyze the results from a quantitative point of view we define an effectiveness factor $\Lambda := \mathscr{J}_\Omega(u_{ini})/\mathscr{J}_\Omega(u_{opt})$, where u_{ini} and u_{opt} are the displacements of the *initial* and *optimized* configurations, respectively. This effectiveness factor can be viewed as a particularization of the standard *geometric advantage (GA)* metric, used in the design and performance studies of compliance mechanisms. See for instance [133]. The *GA* measures the relation between the obtained displacement u_{out} when the device is actuated by an input displacement u_{in}, i.e. $GA := u_{out}/u_{in}$. Here, the factor Λ measures the *GA* with respect to the obtained displacement of the reference (initial) configuration.

7.5.1 Example 1: Amplifier

The first example is the optimization of a displacement amplifier. This device is used to amplify the displacements in a given direction generated by thermal effects. In particular, the design domain considered is presented in Fig. 7.2, in which only one quadrant of the complete domain is represented, based on horizontal and vertical symmetry assumptions (the dashed-dot lines indicate the axes of symmetry). The objective is the maximization of the outward output displacement in the direction e on $\Gamma^\star$ in response to a thermal excitation imposed on Γ_θ. In this case, the boundary condition is given by a linear temperature distribution on Γ_θ, as shown in Fig. 7.2b. The material properties are optimized in white subdomains, while in the light grey regions of Fig. 7.2a, b the material properties are fixed.

In Fig. 7.3, the results for two different volume fractions are shown. In Fig. 7.4 a selected result is shown without the symmetry boundary condition, in which the deformed configuration is represented by the grey region. The variation of the effectiveness values Λ with respect to the final volume fraction are presented in Fig. 7.5. The behavior of the effectiveness factor with respect to the final volume fraction suggests that there exists a volume fraction close to the 40% where the displacement on $\Gamma^\star$ is maximal in the direction e. This value can be interpreted as an *optimal volume fraction* whose associated effectiveness factor is $\Lambda = 4.01$.

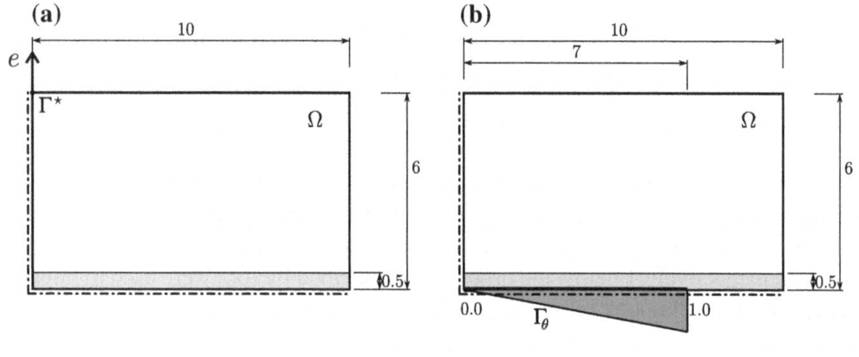

Fig. 7.2 Example 1. Domain and boundary conditions: **a** mechanical and **b** heat problems

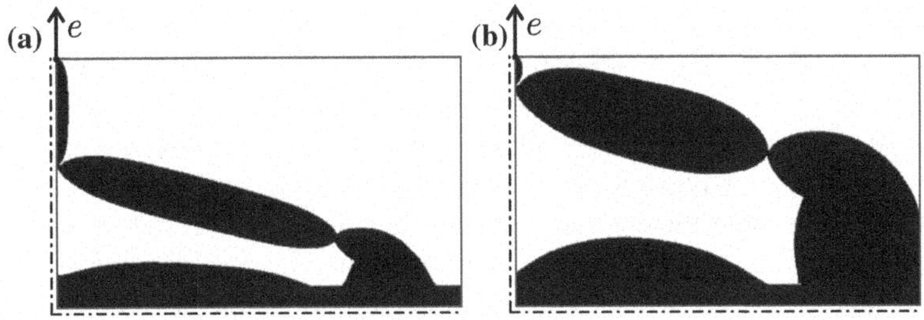

Fig. 7.3 Example 1. Results for different volume fractions: **a** 30% and **b** 50%

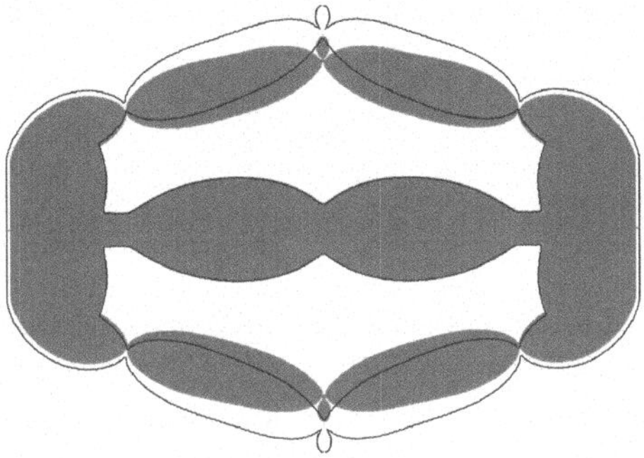

Fig. 7.4 Example 1. Amplifier from Fig. 7.3b

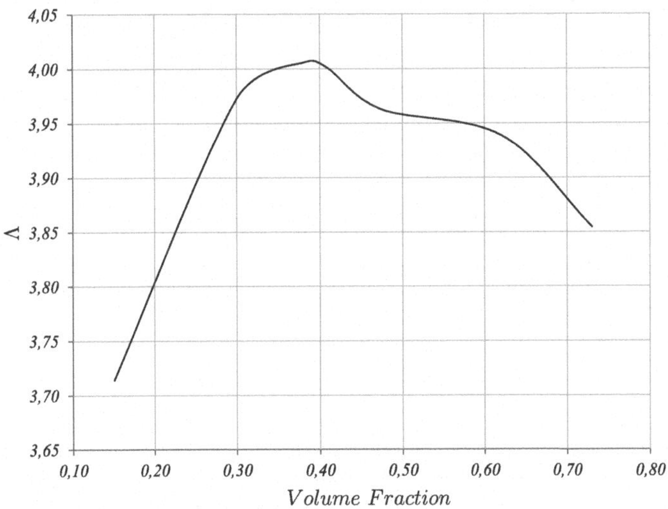

Fig. 7.5 Example 1. Effectiveness factor versus final volume fraction

After an inspection of the obtained results, we note the presence of flexible hinges in the design. These hinges allow for high values of Λ. The hinges generated in optimal design assure easy rotation of the upper lever elements in order to produce maximal displacement along the symmetry axis. Here, we can refer to the simple example in Sect. 7.1.1 illustrating the optimal mode of deformation. However, the hinges are undesirable for obvious reasons. Actually, they can induce large local stress concentration, so in real design the final material segments should be used. Such a pathology is a consequence of the adopted formulation based on compliance maximization. How to avoid these hinges is a subject of recent research. See, for instance, [162, 177]. On the other hand, the results previously presented can be interpreted, from an engineering point of view, as a ring connected by a transversal bar.

7.5.2 Example 2: Inverter with Eccentricity Effect

The second example considers the same domain from the previous experiment, however, the output displacement region $\Gamma^\star$ is changed as depicted in Fig. 7.6. This apparently simple modification in the design domain actually results in a completely different mechanism, since the optimizer seeks an output displacement contrary to the natural movement of the thermomechanical device. In addition, all symmetry assumptions remain valid and the boundary condition for thermal problem is given by a linear temperature distribution on Γ_θ, as shown in Fig. 7.6b. The material properties are optimized in white subdomains, while in the light grey regions of Fig. 7.6a, b the material properties are fixed, as in the previous example.

The results for two different volume fractions are shown in Fig. 7.7. In Fig. 7.8 a selected result is shown without the symmetry boundary condition, whose deformed configuration is represented by the grey region. Here, we also notice the presence of flexible hinges in the design. Referring to Figs. 7.7 and 7.8 it is seen that the

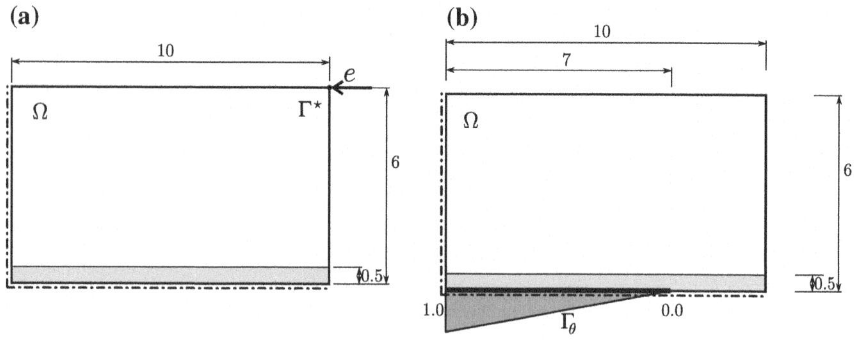

Fig. 7.6 Example 2. Domain and boundary conditions: **a** mechanical and **b** thermal problems

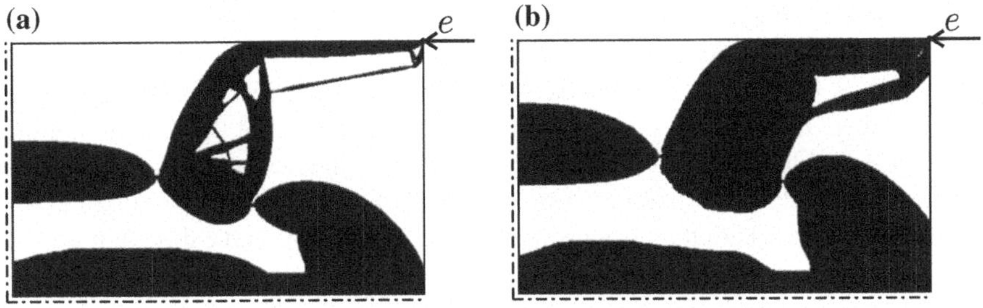

Fig. 7.7 Example 2. Results for different volume fractions: **a** 45% and **b** 60%

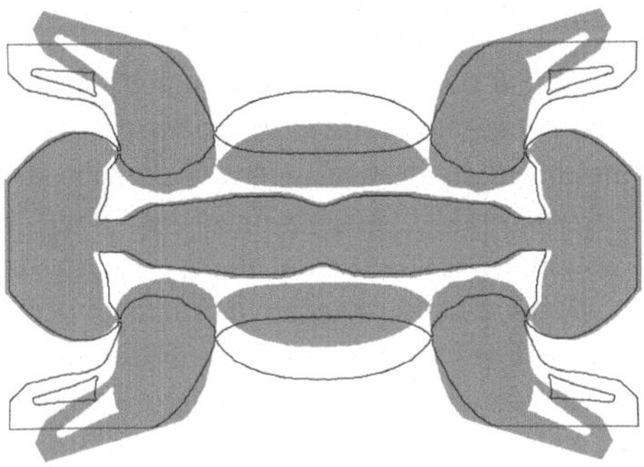

Fig. 7.8 Example 2. Inverter from Fig. 7.7b

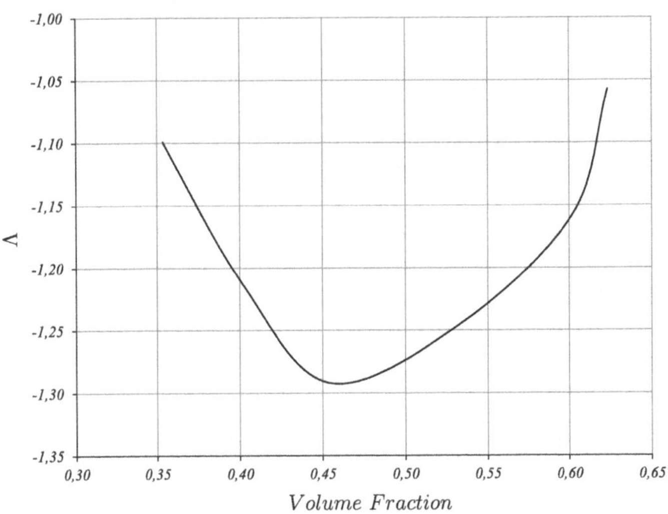

Fig. 7.9 Example 2. Effectiveness factor versus final volume fraction

contact forces at hinge points with upper lever elements induce the rotation moments producing large displacement in the e-direction.

The values for the effectiveness factor Λ for the obtained results are presented in Fig. 7.9, where the negative sign for Λ indicates the inversion of the direction of the displacement. Also, in this example, the effectiveness factor has a minimum value between 45% and 50% of volume fraction. This behavior suggests that there exists a volume fraction where the displacement on $\Gamma^\star$ is maximal in the direction e.

7.6 Final Remarks

In this chapter the topological derivative of the tracking-type shape functional for the semi-coupled thermomechanical model are derived in two spatial dimensions. In order to avoid complicated theoretical derivations such as the ones presented by [111], the thermal expansion coefficients have been fixed. By introducing contrasts on the thermal conductivity coefficient and elastic modulus, the derivations become much simpler, allowing us to focus on the main contribution of this chapter, namely: a simple and analytical expression of the topological derivative to be used in the design of thermal-mechanical actuators, where the contrasts in the material properties are used just to mimic voids. Actually, the information provided by the topological derivative $\mathcal{T}(\widehat{x})$ can be used as a steepest descent direction in an optimal design algorithm. To illustrate this feature, two numerical experiments associated with the topology optimization of actuators have been presented. These simple examples show the applicability of the proposed methodology in the context of optimal design of thermomechanical devices. Furthermore, we have shown that the proposed methodology allows for finding the optimal volume fraction after some realizations. That is, the volume of the actuator which produces the maximal effectiveness factor Λ for a given direction e. Finally, the remarkable simplicity of topological derivative formula (7.72) has to be noted: once the temperature distribution θ (solution of (7.12)), displacement field u (solution of (7.11)), thermal adjoint state φ (solution of (7.18)) and mechanical adjoint state p (solution of (7.17)) are obtained in the original (unperturbed) domain Ω, the topological derivative $\mathcal{T}(\widehat{x})$ can be evaluated at all $\widehat{x} \in \Omega$ using standard postprocessing procedures. Therefore the resulting topology design algorithms based on the topological derivative concept are in general very fast and easy to implement.

Chapter 8
Synthesis of Compliant Piezomechanical Actuators

In control of distributed parameter systems the piezoelectric actuators are of common use. The topology optimization of a multiphysics model in piezoelectricity into two spatial dimensions is considered. The topological derivative of a tracking-type shape functional is derived in its closed form for the purpose of shape optimization of piezoelectric actuators. The optimum design procedure is applied to a micromechanism which transforms the electrical energy supplemented via its piezoceramic part into elastic energy of an actuator. The domain decomposition technique and the Steklov–Poincaré pseudo-differential boundary operator are employed in the asymptotic analysis of the shape functional defined on a part of the boundary of the elastic body under consideration. The new method of sensitivity analysis is general and can be used for the purpose of the shape-topological optimization in a broad class of multiphysics models. The numerical results confirm the efficiency of proposed approach to optimum design in multiphysics. This chapter is based on the paper by Ricardo Amigo, Sebastián Miguel Giusti, Antonio André Novotny, Emílio Carlos Nelli Silva and Jan Sokolowski [10].

8.1 Preliminaries

In this chapter we are interested in the optimal design of piezoelectric actuators, which consist of multi-flexible structures actuated by piezoceramic devices that generate an output displacement in a specified direction on the boundary of the actuated part [235]. The multi-flexible structure transforms the piezoceramic output displacement by amplifying and changing its direction. This kind of mechanism can be manufactured at a very small scale. Therefore, the spectrum of applications of such microtools becomes broader in recent years including microsurgery, nanotechnology processing, cell manipulation, among others. Yet, the development of microtools requires the design of actuated multi-flexible structures which are able to produce complex movements originated from simple expansion/contraction of the piezoceramic actu-

A. A. Novotny et al., *Applications of the Topological Derivative Method*,
Studies in Systems, Decision and Control 188,
https://doi.org/10.1007/978-3-030-05432-8_8

ator. The performance of microtools can be strongly enhanced by optimizing the actuated multi-flexible structures with respect to their shape and their topology. The shape sensitivity analysis of such coupled models has been fully developed in [165] and [166] for quasi-electrostatic layered piezoelectric devices and for non-stationary elastic, piezoelectric and acoustic coupled system, respectively. For the mathematical theory concerning coupled PDEs systems the reader may refer to e.g., [158, 189, 192].

A general approach to deal with shape and topology optimization design is based on the topological derivative. In fact, this relatively new concept represents the first term of the asymptotic expansion of a given shape functional with respect to the small parameter which measures the size of singular domain perturbations, such as holes, inclusions, source-terms and cracks. The topological asymptotic analysis was introduced in the fundamental paper [238] and has been successfully applied in the treatment of problems such as topology optimization [33], inverse analysis [130], image processing [128], multi-scale constitutive modeling [29], fracture mechanics sensitivity analysis [253] and damage evolution modeling [8]. For an account of new developments in this branch of shape optimization we refer to [219].

In particular, the topological derivative is here applied in the context of topology optimization of piezoelectric actuated multi-flexible structures into two spatial dimensions. The basic idea consists in maximizing the performance of the microtool by introducing small inclusions in the multi-flexible elastic part. Since this problem is modeled by a coupled electro-mechanical system, the domain decomposition technique combined with the Steklov–Poincaré pseudo-differential boundary operator is used to derive the first order term of the asymptotic expansion of the shape functional with respect to the small parameter measuring the size of the inclusions. Thus, a new method of topological sensitivity analysis is proposed for the coupled models. In our framework the topological derivatives for the tracking-type functionals are obtained in their closed forms, which can be used e.g., as a steepest descent direction in the microtools design by topology optimization method. However, in shape optimization with the PDEs constraints there are three main issues to be solved, which are:

- Existence of optimal shapes;
- Necessary optimality conditions;
- Numerical methods.

The existence of optimal shapes can be assured by regularity conditions imposed on the boundaries of admissible domains. Without such restrictive constraints on admissible shapes, the existence issue cannot be solved in general. The special structure of the shape optimization problem can be used for the direct proof of the existence, e.g., by the application of Mosco convergence to the elliptic problems. Unfortunately, even in such situation, the obtained results are not in general constructive.

Necessary optimality conditions are obtained by the shape sensitivity analysis. Recently, the asymptotic analysis has been employed in the context of singular boundary perturbations in order to obtain the topological derivatives of shape functionals for elliptic boundary value problems. These kind of results are also obtained for our problem. The knowledge of the shape gradients and the topological derivatives for a specific shape functional is required in order to formulate the necessary optimality conditions as well as to devise numerical methods of shape optimization. There is also the specific structure of shape derivatives obtained for shape differentiable functionals in the form of distributions supported on the moving boundaries [245]. The structure should be taken into account when using the discretization of continuous gradient combined with the standard methods of nonlinear optimization like gradient or Newton methods in the numerical methods.

In order to fix these ideas, let us consider a geometrical domain Ω and its singularly perturbed counterpart $\Omega_\varepsilon = \Omega \setminus \overline{B_\varepsilon}$ obtained by the nucleation of a small hole $B_\varepsilon = \{|x - \widehat{x}| < \varepsilon\}$ with center at an arbitrary point $\widehat{x}$ of Ω. There are two different expansions of

$$\varepsilon \mapsto j(\varepsilon) := J(\Omega_\varepsilon). \tag{8.1}$$

The first one obtained by the classical shape sensitivity analysis for $\varepsilon > 0$, namely

$$j(\varepsilon + \delta) = j(\varepsilon) + \delta j'(\varepsilon) + o(\varepsilon; \delta), \tag{8.2}$$

The second expansion obtained by the asymptotic analysis in singularly perturbed domains for $\varepsilon = 0^+$, that is

$$j(\varepsilon) = j(0) + f(\varepsilon)\mathscr{T}(\widehat{x}) + o(f(\varepsilon)). \tag{8.3}$$

The topological derivative $\widehat{x} \mapsto \mathscr{T}(\widehat{x})$ can be used at the preliminary step of optimization procedure to detect the location and the number of small holes inserted into Ω in order to improve the value of the shape functional. Therefore, the robust formula for the topological derivative is required for the precise performance of this step of the procedure. This goal is achieved in this chapter for the piezo model under consideration by the appropriate adjoint state equation.

It turns out that the nature of Taylor expansion of the shape functional $\varepsilon \mapsto J(\Omega_\varepsilon)$ with respect to $\varepsilon \in [0, \varepsilon_0)$, $\varepsilon_0 > 0$, evaluated in the singularly perturbed domain Ω_ε depends on the boundary conditions of the state equation prescribed on the boundaries $\partial B_\varepsilon = \{|x - \widehat{x}| = \varepsilon\}$. The direct derivation of the one term asymptotic expansion for (8.1) at $\varepsilon = 0^+$ leads to the selfadjoint-extensions of elliptic operators [205]. The appropriate adjoint state combined with the polarization tensor is introduced in order to obtain the appropriate representation of the topological derivative for numerical methods of shape optimization. The latter step in this procedure is complicated for the coupled models. Therefore, in this chapter, a general method is proposed in order to overcome this difficulty following the original ideas presented in [242]. It consists in decomposing the topologically perturbed geometrical domain into subdomains with different physical properties. The mutual influence of the subdomains is affected by the transmission conditions on the interfaces. In other words, the fields of mechanical and electric natures are coupled by the transmission conditions, as well as, by the mathematical models in the interior of each subdomain. This means that in the elastic material of the body the fictitious ring domain $C(R, \varepsilon) := \{\varepsilon < |x - \widehat{x}| < R\}$ is introduced for the purpose of the asymptotic analysis with respect to $\varepsilon \to 0$, and the result of the analysis is expressed on the boundary of the ball $B_R = \{|x - \widehat{x}| < R\}$. From the asymptotic expansions of elastic energy in the interior of B_R or $C(R, \varepsilon)$ the expansion of the Dirichlet-to-Neumann map associated with the ball or with the ring is obtained. Once having the asymptotic expansion in hand, the Dirichlet-to-Neumann operator in B_R is employed as the Steklov–Poincaré boundary pseudodifferential operator in the truncated domain $\Omega_R := \Omega \setminus \overline{B_R}$. In this way, the influence of singularities associated with the limit passage $\varepsilon \to 0$ is modeled in the truncated domain via nonlocal boundary conditions, and the subdomain B_R is eliminated from the analysis. Otherwise, the asymptotic analysis of the coupled model in singularly perturbed geometrical domain should be performed, which would not be the best idea because of the complexity of such an approach. Finally, the domain decomposition method combined with the asymptotic expansions in the ring associated with the small parameter $\varepsilon \to 0$ allows us to find the topological derivative of the shape functional defined in the truncated domain $\Omega \setminus \overline{B_R}$ for all $R > \varepsilon \to 0$. This approach simplifies the topological asymptotic analysis of the shape functional under consideration. See sketch in Fig. 8.1.

Remark 8.1 There is a double notation for the same boundary pseudodifferential operator on ∂B_R depending on the fact whether ∂B_R is considered as the exterior boundary of the fictitious subdomain B_R or the interior boundary of truncated domain Ω_R. More precisely, the nonlocal boundary operator stands for the Steklov–Poincaré operator when acting on the *interior* boundary ∂B_R of the truncated domain Ω_R. Thus, the elasticity boundary value problem in B_R is called the *interior problem of fictitious domain decomposition*, and the coupled model in Ω_R is called the *exterior problem of fictitious domain decomposition* in the notation employed in this chapter. The boundary ∂B_R is used as an interface of the fictitious domain decomposition introduced exclusively for the purpose of asymptotic analysis in the singularly perturbed domain.

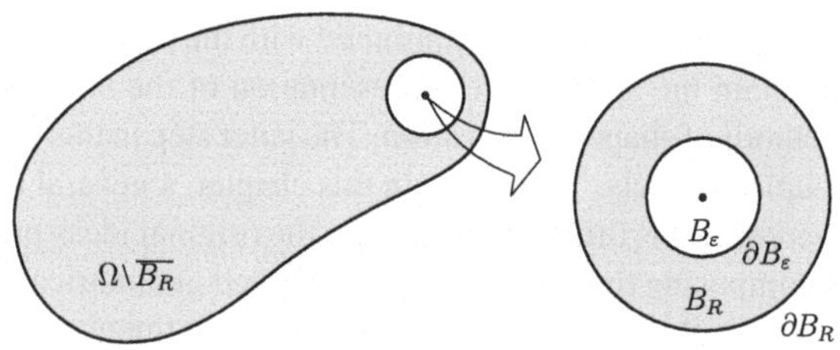

Fig. 8.1 Truncated domain

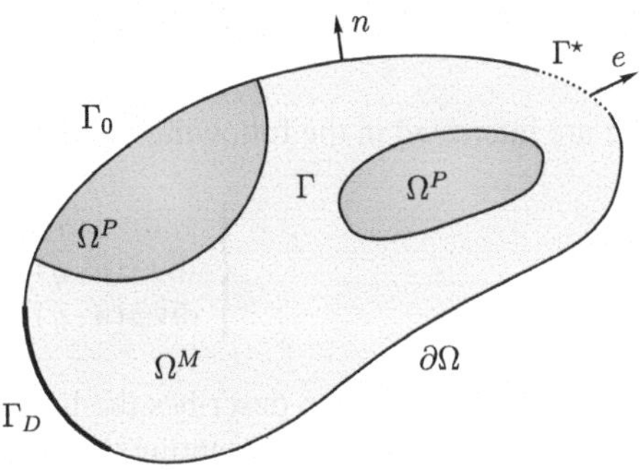

Fig. 8.2 Piezo-elastic coupled problem

8.2 Problem Formulation

Now, we assume that $\Omega = \Omega^M \cup \Gamma \cup \Omega^P$, where the mutually disjoints open domains Ω^P and Ω^M have the common interface Γ, as shown in Fig. 8.2. In our notation, Ω^M and Ω^P represent the regions where mechanical and piezoelectric devices are located, respectively. We consider the coupled model defined in the subdomains Ω^M and Ω^P with different material properties. The models are coupled by the transmission conditions imposed on the interface $\Gamma := \partial\Omega^M \cap \partial\Omega^P$. We are looking for a topology of the mechanical part Ω^M that minimizes a given shape functional defined on $\Gamma^\star \subset \partial\Omega^M$, by keeping the piezoelectric device Ω^P fixed.

In order to define the possible *directions* of shape-topological modifications in the subdomain Ω^M a small inclusion is introduced. It means that we have at our disposal two materials and try to improve the design by an optimal distribution of more expensive material in the mechanical subdomain. The variation of the shape functional associated with the inclusion B_ε is called the *topological derivative* [219, 238]. In order to evaluate the topological derivative we need to recover the asymptotics of solutions to the coupled model, find the asymptotics of the shape functional, and introduce the adjoint state in such a way that the result of asymptotic analysis can be used in numerical methods. This approach is used in [205] for the purpose of asymptotic analysis of the elasticity boundary value problems in three spatial dimensions.

In this work we propose an alternative approach, which is much simpler compared to the general case considered in [205]. Our goal is the identification of the topological derivative in such a way that it can be directly used in numerical methods. Therefore, we combine all the elements of analysis performed in [219, 243] and obtain the method which is sufficiently simple to be used in applied shape-topological design. The method is presented for a coupled model of elastic and piezoelectric materials, however it is general and can be used for more complex models of multiphysics.

8.2.1 The Mechanical Model

We are interested in the following system

$$\begin{cases} \mathrm{div}\sigma(u) = 0 & \text{in } \Omega^M \\ \left.\begin{array}{l} \mathrm{div}S(w,q) = 0 \\ \mathrm{div}\phi(w,q) = 0 \end{array}\right\} & \text{in } \Omega^P \end{cases} \tag{8.4}$$

where the first equation describes the linear elasticity system while the second one is the coupled system representing the electromechanical interaction phenomenon. The equations are coupled on the interface Γ. In particular, $\sigma(u)$ is the mechanical stress tensor, $S(w,q)$ is the electromechanical stress tensor and $\phi(w,q)$ is the electric displacement. The constitutive laws describing the elastic behavior and piezoelectric effects, both in the linearized case of small mechanical deformations and electric fields, are

$$\begin{cases} \sigma(u) = C\nabla^s u \ , \\ S(w,q) = A\nabla^s w + P\nabla q \ , \\ \phi(w,q) = P^T\nabla^s w - K\nabla q \ , \end{cases} \tag{8.5}$$

where u and w are the mechanical and electromechanical displacements, respectively, and q is the electric potential. In addition, C and A are the elasticity fourth-order tensors respectively associated to the elastic and electromechanical parts, P the piezoelectric coupling third-order tensor and K the dielectric second-order tensor. As usual C, A and K satisfy the symmetry conditions $C_{ijkl} = C_{jikl} = C_{klij}$, $A_{ijkl} = A_{jikl} = A_{klij}$, and $K_{ij} = K_{ji}$, whereas P satisfies $P_{ijk} = P_{jik}$. It is assumed for simplicity that all constitutive tensors are piecewise constant, i.e., constant in each sub-domain Ω^M and Ω^P. In the case of isotropic elasticity, the tensor C takes the form

$$C = 2\mu\mathbb{I} + \lambda(I \otimes I) \ , \tag{8.6}$$

where μ and λ are the Lame's coefficients, I and $\mathbb{I}$ are the second and fourth orders identity tensors, respectively. We complement the system (8.4) with the following boundary conditions

$$u = 0 \text{ on } \Gamma_D \ , \quad \sigma(u)n = ku \text{ on } \Gamma^\star \quad \text{and} \quad \begin{cases} q = 0 \text{ on } \Gamma_0 \\ q = \overline{q} \text{ on } \Gamma \end{cases} , \tag{8.7}$$

where $\Gamma^\star$, Γ_D and Γ_0 are parts of the boundary $\partial\Omega$ and n is the outward unit normal vector pointing toward the exterior of Ω. If it is not specified, we consider homogeneous natural (Neumann) boundary conditions of the form $\sigma(u)n = 0$, $S(w,q)n = 0$ and $\phi(w,q) \cdot n = 0$ on a part of $\partial\Omega$. Finally, we consider the following transmission conditions

$$\begin{cases} u = w \\ \sigma(u)n = S(w,q)n \end{cases} \text{on } \Gamma \ , \tag{8.8}$$

where n is the unit normal vector pointing toward the exterior of Ω^M. The variational formulation of the above coupled system reads:

Problem 8.1 Find $\mathrm{u} \in \mathscr{V}$ and $q \in \mathscr{Q}$ such that

$$\begin{cases} \displaystyle\int_{\Omega^M} \sigma(u) \cdot \nabla^s \eta - \int_{\Gamma^\star} ku \cdot \eta + \int_{\Omega^P} S(w, q) \cdot \nabla^s \eta = 0 \quad \forall \eta \in \mathscr{V} , \\ \displaystyle\int_{\Omega^P} \phi(w, q) \cdot \nabla \xi = 0 \quad \forall \xi \in \mathscr{Q}_0 , \end{cases} \tag{8.9}$$

where $\mathrm{u} = u$ in Ω^M and $\mathrm{u} = w$ in Ω^P. The space $\mathscr{V}$ of displacements fields is defined as

$$\mathscr{V} = \left\{ v \in H^1(\Omega) : v|_{\Gamma_D} = 0 \right\} , \tag{8.10}$$

while the electric potentials sets $\mathscr{Q}$ and $\mathscr{Q}_0$ are respectively defined as

$$\mathscr{Q} = \left\{ q \in H^1(\Omega^P) : q|_{\Gamma} = \overline{q},\ q|_{\Gamma_0} = 0 \right\} , \quad \mathscr{Q}_0 = \left\{ q \in H^1(\Omega^P) : q|_{\partial\Omega^P} = 0 \right\} . \tag{8.11}$$

8.2.2 The Shape Functional

We are interested in the topology design of piezoelectric actuators, which consist of multi-flexible structures actuated by piezoceramic devices that generate an output displacement in a specified direction on the boundary of the actuated part. Therefore, let us introduce a tracking-type shape functional of the form

$$J(u) = -\int_{\Gamma^\star} u \cdot e . \tag{8.12}$$

where e is used to denote a given direction on the boundary $\Gamma^\star$ (see Fig. 8.2).

8.2.3 The Adjoint State

As mentioned before, we are going to evaluate the shape gradient of functional (8.12). For further simplifications, we introduce an adjoint system of the form

$$\begin{cases} \operatorname{div} \sigma(u^a) = 0 \quad \text{in } \Omega^M \\ \left. \begin{array}{l} \operatorname{div} S^a(w^a, q^a) = 0 \\ \operatorname{div} \phi^a(w^a, q^a) = 0 \end{array} \right\} \text{in } \Omega^P \end{cases} , \tag{8.13}$$

where $\sigma(u^a)$, $S^a(w^a, q^a)$ and $\phi^a(w^a, q^a)$, respectively are the adjoint mechanical stress tensor, electromechanical stress tensor and electrical displacement, given by

$$\begin{cases} \sigma(u^a) = C\nabla^s u^a \,, \\ S^a(w^a, q^a) = A\nabla^s w^a - P\nabla q^a \,, \\ \phi^a(w^a, q^a) = -P^T\nabla^s w^a - K\nabla q^a \,, \end{cases} \tag{8.14}$$

where u^a, w^a and q^a are the adjoint mechanical displacement, electromechanical displacement and electric potential, respectively. The system (8.13) has the following boundary conditions

$$u^a = 0 \text{ on } \Gamma_D, \qquad \sigma(u^a)n = ku^a + e \text{ on } \Gamma^\star \quad \text{and} \quad q^a = 0 \text{ on } \partial\Omega^P \tag{8.15}$$

and the transmission conditions

$$\begin{cases} u^a = w^a \\ \sigma(u^a)n = S^a(w^a, q^a)n \end{cases} \text{on } \Gamma \,. \tag{8.16}$$

The variational formulation of the coupled system for adjoint state equations reads:

Problem 8.2 Find $\mathrm{u}^a \in \mathscr{V}$ and $q^a \in \mathscr{Q}_0$ such that

$$\begin{cases} \int_{\Omega^M} \sigma(u^a)\cdot\nabla^s\eta - \int_{\Gamma^\star} ku^a\cdot\eta + \int_{\Omega^P} S^a(w^a, q^a)\cdot\nabla^s\eta = \int_{\Gamma^\star} e\cdot\eta \quad \forall\eta \in \mathscr{V} \,, \\ \int_{\Omega^P} \phi^a(w^a, q^a)\cdot\nabla\xi = 0 \qquad \forall\xi \in \mathscr{Q}_0 \,, \end{cases} \tag{8.17}$$

where $\mathrm{u}^a = u^a$ in Ω^M and $\mathrm{u}^a = w^a$ in Ω^P.

8.3 Topological Derivative

The topological derivative of functional (8.12) is evaluated for the insertion of a small inclusion in Ω^M with the material properties depending on the contrast. To describe the topological perturbation of Ω^M we introduce a piecewise constant function γ_ε of the form

$$\gamma_\varepsilon = \gamma_\varepsilon(x) := \begin{cases} 1 \text{ if } x \in \Omega^M \setminus \overline{B_\varepsilon} \\ \gamma \text{ if } x \in B_\varepsilon \end{cases} \,, \tag{8.18}$$

where $0 < \gamma < \infty$ is the contrast parameter on the material properties and $B_\varepsilon(\widehat{x})$ for $\widehat{x} \in \Omega^M$. Note that in this case the topologies of the original and perturbed domains are preserved. However, we are introducing a non-smooth perturbation in the coefficients of the differential operator through the contrast γ_ε, by changing the material property of the background in a small region $B_\varepsilon \subset \Omega^M$. Therefore, the sensitivity of the shape functional with respect to the nucleation of an inclusion can also be handle through the topological asymptotic analysis concept, which actually is the best approach for such a problem. The variational formulation associated with the perturbed coupled system reads:

Problem 8.3 Find $\mathrm{u}_\varepsilon \in \mathscr{V}$ and $q_\varepsilon \in \mathscr{Q}$ such that

$$\begin{cases} \int_{\Omega^M} \sigma_\varepsilon(u_\varepsilon) \cdot \nabla^s \eta - \int_{\Gamma^\star} k u_\varepsilon \cdot \eta + \int_{\Omega^P} S(w_\varepsilon, q_\varepsilon) \cdot \nabla^s \eta = 0 & \forall \eta \in \mathscr{V} , \\ \int_{\Omega^P} \phi(w_\varepsilon, q_\varepsilon) \cdot \nabla \xi = 0 & \forall \xi \in \mathscr{Q}_0 , \end{cases} \tag{8.19}$$

where $\sigma_\varepsilon(u_\varepsilon) = \gamma_\varepsilon C \nabla^s u_\varepsilon$. In addition, $\mathrm{u}_\varepsilon = u_\varepsilon$ in Ω^M and $\mathrm{u}_\varepsilon = w_\varepsilon$ in Ω^P.

8.3.1 Abstract Setting of the Problem

In this chapter a coupled model is considered in the domain Ω. The coupled system (8.9) is well-posed, and can be written in the strong form as an abstract equation for the unknown functions $U := (u, w, q)$,

$$LU = F \tag{8.20}$$

in the appropriate function spaces over the domain Ω, where F is a vector that has the generalized loading system for the problem and L is the matrix of the complete system of equations. The weak form reads

$$\mathscr{L}(U, \Phi) = (F, \Phi) , \tag{8.21}$$

with the test functions Φ. The bilinear form associated with the elastic component of the coupled model in the subdomain Ω^M is simply given by standard expression of linear elasticity

$$(u, \eta) \mapsto a(\Omega^M; u, \eta) := \int_{\Omega^M} \sigma(u) \cdot \nabla^s \eta \tag{8.22}$$

in the unperturbed subdomain Ω^M, as well as by

$$(u, \eta) \mapsto a_\varepsilon(\Omega^M; u, \eta) := \int_{\Omega^M} \sigma_\varepsilon(u) \cdot \nabla^s \eta \tag{8.23}$$

in the perturbed subdomain by an inclusion. Here $\varepsilon \to 0$ is the parameter which governs the size of the topological perturbation. In the latter case, the weak formulation of the coupled model also depends on the small parameter ε, and can be rewritten as follows

$$\mathscr{L}_\varepsilon(U_\varepsilon, \Phi) = (F, \Phi) , \tag{8.24}$$

or written in the strong form $L_\varepsilon U_\varepsilon = F$. The perturbed system is also well-posed for $\varepsilon \in [0, \varepsilon_0)$, with $\varepsilon_0 > 0$, i.e., the inverse operator is uniformly bounded: $\|U_\varepsilon\| \leq$

$C\|F\|$ in appropriate norms. However, in the case of a cavity the associated function spaces are obviously dependent on the small parameter ε. The case of an inclusion is therefore the regular perturbation of the problem in the fixed function spaces setting.

We proceed further with the domain decomposition technique in the subdomain $\Omega^M := B_R \cup \Gamma_R \cup \Omega_R^M$, where B_R is the ball of radius R which contains the topological perturbation denoted by B_ε, $\Gamma_R \equiv \partial B_R$ stands for the boundary of B_R, and the remaining subdomain $\Omega_R^M := \Omega^M \setminus \overline{B_R}$ is far from the singular topological domain perturbation.

Let u_ε denote the solution of coupled equations in the perturbed domain Ω i.e., including the inclusion B_ε. We are going to show, that the restriction of u_ε to the truncated domain solves a boundary value problem with the nonlocal boundary conditions on Γ_R defined by the Steklov–Poincaré operator. Since for all $\varepsilon \in [0, \varepsilon_0)$,

$$a_\varepsilon(\Omega^M; u_\varepsilon, \eta) = a(\Omega_R^M; u_\varepsilon, \eta) + a_\varepsilon(B_R; u_\varepsilon, \eta) \tag{8.25}$$

and the last term is equivalent to the Steklov–Poincaré component by construction

$$a_\varepsilon(B_R; u_\varepsilon, \eta) \equiv (\mathscr{A}_\varepsilon(u_\varepsilon), \eta)_{\Gamma_R} \,. \tag{8.26}$$

Henceforth the bilinear form in the topologically perturbed domain Ω^M can be replaced by the bilinear form in the unperturbed domain Ω_R^M, however with a nonlocal pseudo-differential operator

$$a_\varepsilon(\Omega^M; u_\varepsilon, \eta) = a(\Omega_R^M; u_\varepsilon, \eta) + (\mathscr{A}_\varepsilon(u_\varepsilon), \eta)_{\Gamma_R} \,. \tag{8.27}$$

This replacement is in fact crucial for the proofs of topological differentiability for the shape functionals defined for the coupled system. Actually, the asymptotic expansion of the solutions to the coupled model in the truncated domain can be easily deduced from the well-posedness of the model. It means that for the linear model, and the regular perturbations of the differential operator, the asymptotic expansion of the Steklov–Poincaré operator

$$\mathscr{A}_\varepsilon = \mathscr{A} + f(\varepsilon)\mathscr{A}' + \mathscr{R}_\varepsilon, \tag{8.28}$$

with $f(\varepsilon) \sim |B_\varepsilon|$, implies the asymptotic expansion of the solutions to the coupled model in the truncated domain $\Omega_R^M \cup \Gamma_R \cup \Omega^P$ of the same form

$$U_\varepsilon = U + f(\varepsilon)U' + \widetilde{U}_\varepsilon \,, \tag{8.29}$$

where $\widetilde{U}_\varepsilon$ is the remainder, namely $||\widetilde{U}_\varepsilon|| = o(f(\varepsilon))$ in appropriated norms.

8.3.2 Topological Asymptotic Expansion of the Steklov–Poincaré Operator

The proposed method of asymptotic analysis is employed now to the coupled system defined in Ω_R. The dependence of the model on the small parameter $\varepsilon \to 0$ occurs in the nonlocal boundary conditions imposed on Γ_R. The variational form of (8.19) restricted to Ω_R is obtained,

$$\left\{\begin{aligned} &\int_{\Omega^M\setminus\overline{B_R}} \sigma_\varepsilon(u_\varepsilon)\cdot\nabla^s\eta + \int_{\partial B_R} \mathscr{A}_\varepsilon(u_\varepsilon)\cdot\eta - \\ &\quad\int_{\Gamma^\star} k u_\varepsilon\cdot\eta + \int_{\Omega^P} S(w_\varepsilon, q_\varepsilon)\cdot\nabla^s\eta = 0 \quad \forall\eta\in\mathscr{V}\,, \\ &\qquad\qquad\int_{\Omega^P} \phi(w_\varepsilon, q_\varepsilon)\cdot\nabla\xi = 0 \quad \forall\xi\in\mathscr{Q}_0\,, \end{aligned}\right. \tag{8.30}$$

where $B_R = B_R(\widehat{x})$ is a ball of radius R and center at $\widehat{x}\in\Omega^M$.

The Steklov–Poincaré operator on the interior boundary ∂B_R of the truncated domain Ω_R

$$\mathscr{A}_\varepsilon : \varphi\in H^{1/2}(\partial B_R)\to\sigma_\varepsilon(v_\varepsilon)n\in H^{-1/2}(\partial B_R)\,, \tag{8.31}$$

by construction coincides with the Dirichlet-to-Neumann map of the linear elasticity on the ball B_R,

$$\left\{\begin{aligned} \operatorname{div}\sigma_\varepsilon(v_\varepsilon) &= 0 && \text{in } B_R\,, \\ \sigma_\varepsilon(v_\varepsilon) &= \gamma_\varepsilon C\nabla^s v_\varepsilon\,, && \\ v_\varepsilon &= \varphi && \text{on } \partial B_R\,, \\ \left.\begin{aligned}[\![v_\varepsilon]\!] &= 0 \\ [\![\sigma_\varepsilon(v_\varepsilon)]\!]n &= 0\end{aligned}\right\} & && \text{on } \partial B_\varepsilon\,, \end{aligned}\right. \tag{8.32}$$

with $\mathscr{A}_\varepsilon(\varphi) = \sigma_\varepsilon(v_\varepsilon)n$, which assures the identity $v_\varepsilon = u_\varepsilon|_{B_R}$ in B_R, where u_ε is the solution of the perturbed problem in Ω.

Remark 8.2 If the Steklov–Poincaré operator $\varphi\mapsto\mathscr{A}_\varepsilon(\varphi)$ of problem (8.30) is the Dirichlet-to-Neumann map defined by (8.32), then the solution to (8.30) coincides with the restriction to Ω_R of the solution to perturbed problem in Ω.

The identity for the energy functional of (8.32) holds

$$\begin{aligned} 0 = -\int_{B_R} \operatorname{div}\sigma_\varepsilon(v_\varepsilon)\cdot v_\varepsilon &= \int_{B_R} \sigma_\varepsilon(v_\varepsilon)\cdot\nabla^s v_\varepsilon - \int_{\partial B_R} \sigma_\varepsilon(v_\varepsilon)n\cdot v_\varepsilon \\ &= \int_{B_R} \sigma_\varepsilon(v_\varepsilon)\cdot\nabla^s v_\varepsilon - \int_{\partial B_R} \mathscr{A}_\varepsilon(\varphi)\cdot\varphi\,, \end{aligned} \tag{8.33}$$

hence the elastic energy in B_R equals to the energy of the Steklov–Poincaré operator on the boundary. Thus we conclude that the asymptotic expansion of the Steklov–

Poincaré operator on the common boundary ∂B_R equals to the asymptotic expansion of the elastic energy in the domain B_R. Namely

$$\int_{B_R} \sigma_\varepsilon(v_\varepsilon) \cdot \nabla^s v_\varepsilon = \int_{\partial B_R} \mathscr{A}_\varepsilon(\varphi) \cdot \varphi \,, \tag{8.34}$$

for the mapping defined by (8.32)

$$\varphi \in H^{1/2}(\partial B_R) \to \sigma_\varepsilon(v_\varepsilon) n \in H^{-1/2}(\partial B_R) \,. \tag{8.35}$$

Since the operator $\mathscr{A}_\varepsilon$ is symmetric, we can also write

$$\int_{B_R} \sigma_\varepsilon(v_\varepsilon) \cdot \nabla^s v_\varepsilon = \langle \mathscr{A}_\varepsilon(\varphi), \varphi \rangle_{(H^{-1/2} \times H^{1/2})(\partial B_R)} \,. \tag{8.36}$$

It is well-know that the topological asymptotic expansion for the energy functional takes the following form [219]:

$$\int_{B_R} \sigma_\varepsilon(v_\varepsilon) \cdot \nabla^s v_\varepsilon = \int_{B_R} \sigma(v) \cdot \nabla^s v + f(\varepsilon) \mathbb{P}_\gamma \sigma(v(\widehat{x})) \cdot \nabla^s v(\widehat{x}) + o(f(\varepsilon)) \,, \tag{8.37}$$

where $v = u|_{B_R}$ is the solution to the original (unperturbed) problem (8.9) and $\mathbb{P}_\gamma$ is the Pólya-Szegö polarization tensor [16]. According to the expansion 8.28 and by symmetry of the Steklov–Poincaré operator, the expansion of the energy functional can also be written as

$$\langle \mathscr{A}_\varepsilon(\varphi), \vartheta \rangle = \langle \mathscr{A}(\varphi), \vartheta \rangle + f(\varepsilon) \langle \mathscr{B}(\varphi), \vartheta \rangle + \langle \mathscr{R}_\varepsilon(\varphi), \vartheta \rangle \,, \tag{8.38}$$

where $\mathscr{B}(\varphi) := \mathscr{A}'(\varphi)$ and $\langle \mathscr{R}_\varepsilon(\varphi), \vartheta \rangle = o(f(\varepsilon))$. Then, from the asymptotic expansion of the energy functional, we get

$$\langle \mathscr{B}(\varphi), \vartheta \rangle = \mathbb{P}_\gamma \sigma(\varphi(\widehat{x})) \cdot \nabla^s \vartheta(\widehat{x}) \quad \forall \widehat{x} \in \Omega^M \,. \tag{8.39}$$

8.3.3 Topological Asymptotic Expansion of the Solution

We consider the following ansätze for the solutions $u_\varepsilon, w_\varepsilon, q_\varepsilon$ to the topologically perturbed coupled system (8.19)

$$u_\varepsilon = u + f(\varepsilon) g + \widetilde{u}_\varepsilon \,, \tag{8.40}$$

$$w_\varepsilon = w + f(\varepsilon) h + \widetilde{w}_\varepsilon \,, \tag{8.41}$$

$$q_\varepsilon = q + f(\varepsilon) p + \widetilde{q}_\varepsilon \,, \tag{8.42}$$

where u, w, q are solutions to the original (unperturbed) coupled system (8.9); g, h, p are the first order asymptotic correction terms and $\widetilde{u}_\varepsilon$, $\widetilde{w}_\varepsilon$, $\widetilde{q}_\varepsilon$ are the remainders. Now, we plug these ansätze in (8.30) and collect the terms with the same powers of ε to obtain three boundary value problems. The first problem for u, w and q

$$\begin{cases} \displaystyle\int_{\Omega^M\setminus\overline{B_R}} \sigma(u)\cdot\nabla^s\eta + \int_{\partial B_R} \mathscr{A}(u)\cdot\eta - \\ \displaystyle\qquad \int_{\Gamma^\star} ku\cdot\eta + \int_{\Omega^P} S(w,q)\cdot\nabla^s\eta = 0 \quad \forall\eta\in\mathscr{V}\,, \\ \displaystyle\qquad\qquad \int_{\Omega^P} \phi(w,q)\cdot\nabla\xi = 0 \quad \forall\xi\in\mathscr{Q}_0\,. \end{cases} \tag{8.43}$$

The second problem for g, h and p

$$\begin{cases} \displaystyle\int_{\Omega^M\setminus\overline{B_R}} \sigma(g)\cdot\nabla^s\eta + \int_{\partial B_R} (\mathscr{A}(g)+\mathscr{B}(u))\cdot\eta - \\ \displaystyle\qquad \int_{\Gamma^\star} kg\cdot\eta + \int_{\Omega^P} S(h,p)\cdot\nabla^s\eta = 0 \quad \forall\eta\in\mathscr{V}\,, \\ \displaystyle\qquad\qquad \int_{\Omega^P} \phi(h,p)\cdot\nabla\xi = 0 \quad \forall\xi\in\mathscr{Q}_0\,, \end{cases} \tag{8.44}$$

and the third problem for the remainders $\widetilde{u}_\varepsilon$, $\widetilde{w}_\varepsilon$ and $\widetilde{q}_\varepsilon$

$$\begin{cases} \displaystyle\int_{\Omega^M\setminus\overline{B_R}} \sigma_\varepsilon(\widetilde{u}_\varepsilon)\cdot\nabla^s\eta + \int_{\partial B_R} \mathscr{A}_\varepsilon(\widetilde{u}_\varepsilon)\cdot\eta - \\ \displaystyle\quad \int_{\Gamma^\star} k\widetilde{u}_\varepsilon\cdot\eta + \int_{\Omega^P} S(\widetilde{w}_\varepsilon,\widetilde{q}_\varepsilon)\cdot\nabla^s\eta = \int_{\partial B_R} \mathfrak{F}_\varepsilon\cdot\eta \quad \forall\eta\in\mathscr{V}\,, \\ \displaystyle\qquad \int_{\Omega^P} \phi(\widetilde{w}_\varepsilon,\widetilde{q}_\varepsilon)\cdot\nabla\xi = 0 \qquad\qquad \forall\xi\in\mathscr{Q}_0\,, \end{cases} \tag{8.45}$$

where the source $\mathfrak{F}_\varepsilon$ is given by

$$\mathfrak{F}_\varepsilon = -(\mathscr{R}_\varepsilon(u) + f(\varepsilon)\mathscr{R}_\varepsilon(g) + f(\varepsilon)^2\mathscr{B}(g))\,. \tag{8.46}$$

The estimations $||\widetilde{u}_\varepsilon||_{H^1(\Omega^M\setminus\overline{B_R})} = o(f(\varepsilon))$, $||\widetilde{w}_\varepsilon||_{H^1(\Omega^P)} = o(f(\varepsilon))$ and $||\widetilde{q}_\varepsilon||_{H^1(\Omega^P)} = o(f(\varepsilon))$ hold true for the remainders [219].

8.3.4 Topological Asymptotic Expansion of the Shape Functional

Now we are in position to establish the asymptotic expansion of the shape functional and obtain its topological derivative. After introducing the first ansätz in the shape functional associated to the perturbed problem, we have

$$
\begin{aligned}
J(u_\varepsilon) &= -\int_{\Gamma^\star} (u + f(\varepsilon) g + \widetilde{u}_\varepsilon) \cdot e \\
&= -\int_{\Gamma^\star} u \cdot e - f(\varepsilon) \int_{\Gamma^\star} g \cdot e - \int_{\Gamma^\star} \widetilde{u}_\varepsilon \cdot e \\
&= J(u) - f(\varepsilon) \int_{\Gamma^\star} g \cdot e + o(f(\varepsilon)) \, .
\end{aligned} \tag{8.47}
$$

Now, let us rewrite the adjoint system (8.17) as

$$
\left\{
\begin{array}{l}
\displaystyle\int_{\Omega^M \setminus \overline{B_R}} \sigma(u^a) \cdot \nabla^s \eta + \int_{\partial B_R} \mathscr{A}(u^a) \cdot \eta - \\
\qquad \displaystyle\int_{\Gamma^\star} k u^a \cdot \eta + \int_{\Omega^P} S^a(w^a, q^a) \cdot \nabla^s \eta = \int_{\Gamma^\star} e \cdot \eta \quad \forall \eta \in \mathscr{V} \, , \\
\qquad\qquad \displaystyle\int_{\Omega^P} \phi^a(w^a, q^a) \cdot \nabla \xi = 0 \qquad\qquad \forall \xi \in \mathscr{Q}_0 \, .
\end{array}
\right. \tag{8.48}
$$

By taking g, h and p as the test functions in (8.48) we have the following equalities

$$
\int_{\Omega^M \setminus \overline{B_R}} \sigma(u^a) \cdot \nabla^s g + \int_{\partial B_R} \mathscr{A}(u^a) \cdot g - \int_{\Gamma^\star} k u^a \cdot g +
$$

$$
\int_{\Omega^P} A \nabla^s w^a \cdot \nabla^s h - \int_{\Omega^P} P \nabla q^a \cdot \nabla^s h = \int_{\Gamma^\star} e \cdot g \, , \tag{8.49}
$$

$$
-\int_{\Omega^P} P^\top \nabla^s w^a \cdot \nabla p - \int_{\Omega^P} K \nabla q^a \cdot \nabla p = 0 \, . \tag{8.50}
$$

On the other hand, by taking u^a, w^a and q^a as the test functions in (8.44) we obtain

$$
\int_{\Omega^M \setminus \overline{B_R}} \sigma(g) \cdot \nabla^s u^a + \int_{\partial B_R} (\mathscr{A}(g) + \mathscr{B}(u)) \cdot u^a - \int_{\Gamma^\star} k g \cdot u^a +
$$

$$
\int_{\Omega^P} A \nabla^s h \cdot \nabla^s w^a + \int_{\Omega^P} P \nabla p \cdot \nabla^s w^a = 0 \, , \tag{8.51}
$$

$$
\int_{\Omega^P} P^\top \nabla^s h \cdot \nabla q^a - \int_{\Omega^P} K \nabla p \cdot \nabla q^a = 0 \, . \tag{8.52}
$$

Combining the above equalities yields the following important result

$$
\begin{aligned}
\int_{\Gamma^\star} e \cdot g &= -\int_{\partial B_R} \mathscr{B}(u) \cdot u^a \\
&= -\langle \mathscr{B}(u), u^a \rangle_{(H^{-1/2} \times H^{1/2})(\partial B_R)} \\
&= -\mathbb{P}_\gamma \sigma(u(\widehat{x})) \cdot \nabla^s u^a(\widehat{x}) \, ,
\end{aligned} \tag{8.53}
$$

where we have considered the symmetry of the bilinear forms. Finally, the topological asymptotic expansion of the shape functional leads to

$$J(u_\varepsilon) = J(u) + f(\varepsilon)\mathbb{P}_\gamma \sigma(u(\widehat{x})) \cdot \nabla^s u^a(\widehat{x}) + o(f(\varepsilon)) . \tag{8.54}$$

By assuming that the inclusion is far from the piezoelectric part, the topological derivative is given by the following closed formula

$$\mathcal{T}(\widehat{x}) = \mathbb{P}_\gamma \sigma(u(\widehat{x})) \cdot \nabla^s u^a(\widehat{x}) \quad \forall \widehat{x} \in \Omega^M . \tag{8.55}$$

where u and u^a are solutions to the original unperturbed direct (8.9) and adjoint (8.17) systems, respectively.

Corollary 8.1 *In two spatial dimensions, the function $f(\varepsilon) = \varepsilon^2$ and the polarization tensor for inclusions ($0 < \gamma < \infty$) reads [219] (see also [77])*

$$\mathbb{P}_\gamma = \pi \frac{1-\gamma}{1+\gamma\beta}\left((1+\beta)\mathbb{I} + \frac{1}{2}(\alpha-\beta)\frac{1-\gamma}{1+\gamma\alpha} I \otimes I\right) . \tag{8.56}$$

with the constants α and β given by

$$\alpha = \frac{\lambda+\mu}{\mu} \quad \text{and} \quad \beta = \frac{\lambda+3\mu}{\lambda+\mu} . \tag{8.57}$$

In three spatial dimensions, the function $f(\varepsilon) = \varepsilon^3$ and the polarization tensor for holes ($\gamma = 0$) yields [219] (see also [105])

$$\mathbb{P}_0 = 2\pi \frac{1-\nu}{7-5\nu}\left(10\mathbb{I} - \frac{1-5\nu}{1-2\nu} I \otimes I\right) , \tag{8.58}$$

where ν is the Poisson ratio.

8.4 Numerical Experiments

The topology design algorithm developed by [25] is adopted in order to solve the optimization problem, which is based on the topological derivative concept together with a level-set domain representation method. See Algorithm 1 presented in Chap. 6. In all examples, the materials considered are PZT-5A in fixed piezoelectric domains and Nickel or Titanium in optimizable metallic domains. The material properties of PZT-5A are given in Table 8.1. Nickel is characterized by $\rho = 8.908\,(\mathrm{g/cm^3})$, $E = 204.676\,(\mathrm{GPa})$ and $\nu = 0.287$, while Titanium by $\rho = 4.507\,(\mathrm{g/cm^3})$, $E = 109.403\,(\mathrm{GPa})$ and $\nu = 0.287$. Finally, the algorithmic parameters are set as $\gamma^* = 1 \times 10^{-3}$, $\varepsilon_\theta = 1^o$ and $\varepsilon_\kappa = 1 \times 10^{-4}$, while ρ is targeted according to the required final volume M.

Table 8.1 PZT-5A properties

ρ	c_{11}	c_{12}	c_{13}	c_{33}	c_{44}	e_{13}	e_{15}	e_{33}	ε_{11}	ε_{33}
6.080	137.0	69.7	71.6	124.0	31.4	−4.0	10.4	13.8	7.9473	5.1507
g/cm^3	GPa	GPa	GPa	GPa	GPa	C/m^2	C/m^2	C/m^2	nF/m	nF/m

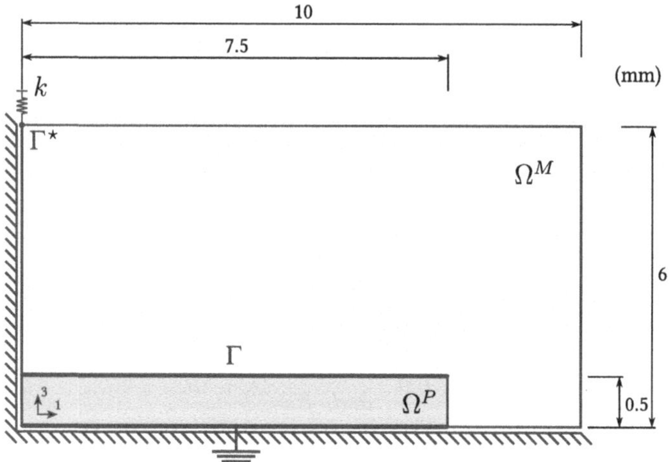

Fig. 8.3 Example 1: Design domain for piezoelectric moonie (dimensions in mm)

8.4.1 Example 1: Moonie

The first implementation example is the optimization of a moonie, which is a widespread device used to amplify the displacements generated by piezoelectric ceramics. In particular, the design domain considered is shown in Fig. 8.3, in which only one quadrant of the complete domain is represented, based on horizontal and vertical symmetry assumptions. The objective is the maximization of the outward output displacement in the region $\Gamma^{\star}$ in response to some electric potential imposed to the electrode Γ, from the fixed ceramic domain denoted by Ω^P. In the output region $\Gamma^{\star}$, a spring k ensures enough stiffness to the resulting topology obtained within the metallic design domain Ω^s.

In following results, the domain is discretized with a regular mesh of 1920 linear triangles elements at the beginning of the optimization procedure and 491520 elements at the end. The electric potential applied to Γ is -100 V. The obtained results considering different spring stiffness k is presented in Fig. 8.4. The deformed configuration of a selected result is presented in Fig. 8.5. As the stiffness value increases, the coupling structure becomes stiffer and the hinges disappear.

The method gives a quite clear topology. We notice the presence of flexible hinges in the design which can be avoided by using a stress constrained formulation [177].

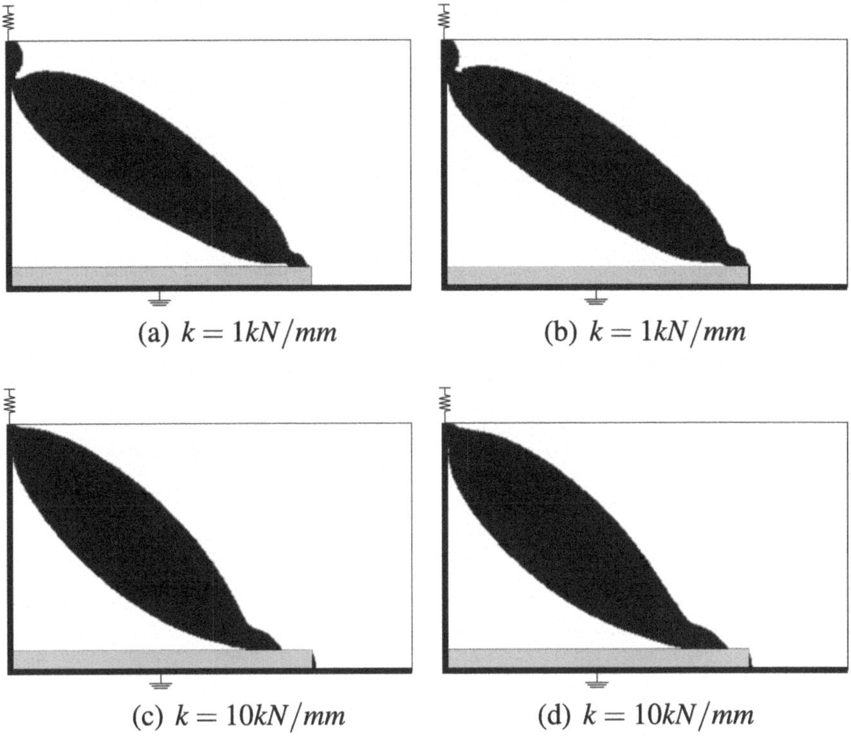

Fig. 8.4 Example 1. Results for Nickel (left) and Titanium (right) for $VF = 0.30$ and different spring stiffness k

Fig. 8.5 Example 1: Deformed configuration for result from Fig. 8.4d

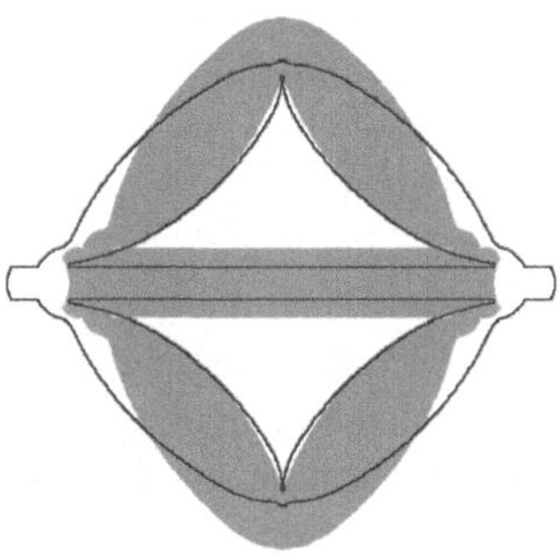

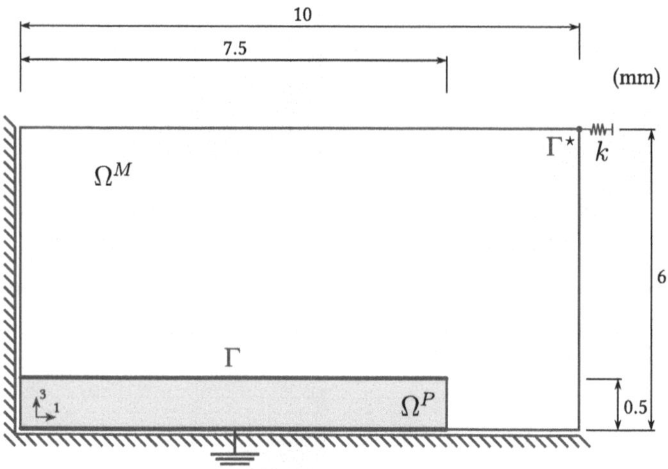

Fig. 8.6 Example 2: Design domain for piezoelectric inverter (dimensions in mm)

However, from the deformed configurations we noticed that the actuator generates the desired displacement.

8.4.2 Example 2: Inverter

The second example considers the same domain from previous section, however, the output displacement region $\Gamma^{\star}$ is changed, as depicted in Fig. 8.6. This apparently simple modification in the design domain actually implies in a completely different mechanism, since it seeks an output displacement contrary to the natural movement of the structure, and this is the reason is called inverter. The design domain remains being Ω^{s}, the objective is still the maximization of the outward output displacement in $\Gamma^{\star}$ and all symmetry assumptions are also valid.

The domain is again discretized with the same mesh of the previous example, the electric potential applied to Γ is -100 V and the spring stiffness is set to 1 kN/mm. In Fig. 8.7, it is shown the results for two different volume constraints by considering Nickel and Titanium design domains. The deformed configuration of a selected result is shown in Fig. 8.8.

The presence of flexible hinges is also noticed in the results. Besides, results from Fig. 8.7a, b show that the method is clearly able to obtain the topology and not only the external shape of the coupling structure. From the deformed configuration we noticed that the inverter generates a displacement opposite to its natural behavior.

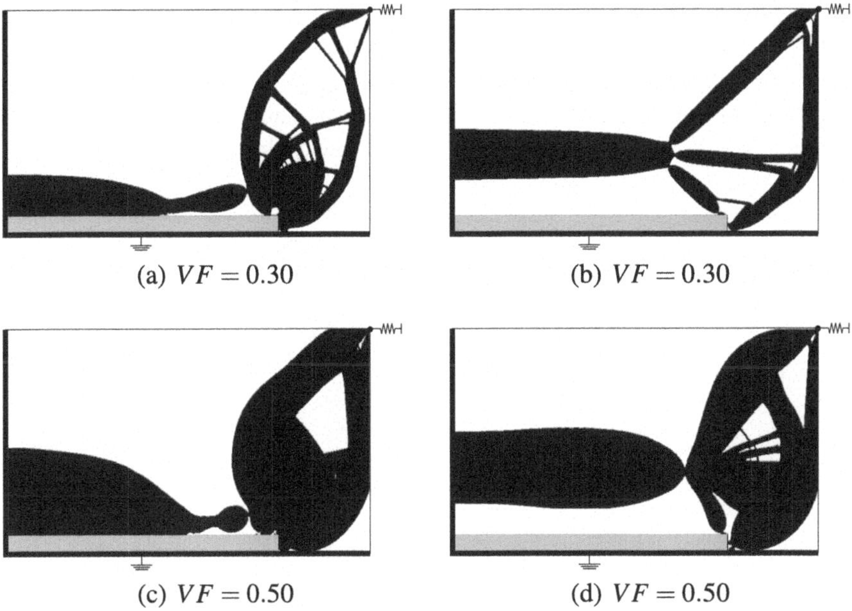

Fig. 8.7 Example 2: Results for Nickel (left) and Titanium (right) for different volume constraints VF

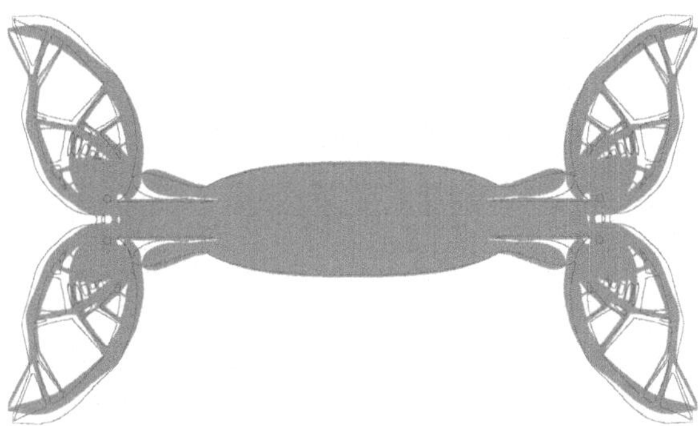

Fig. 8.8 Example 2: Deformed configuration for result from Fig. 8.7a

8.5 Final Remarks

In this chapter the topological derivatives of the tracking-type shape functional for the coupled models of elasto-piezoelectric type are derived in two and three spatial dimensions. The associated shape optimization problems are already analyzed from the point of view of shape optimization in the former papers [165, 166]. In this chapter the preceding results are completed by the topological asymptotic analysis. The remarkable simplicity of the closed form sensitivity given by (8.55) is to be noted. In fact, once the solutions u, w, q and u^a, w^a, q^a to the original (unperturbed) direct (8.9) and adjoint (8.17) coupled systems have been obtained, the topological derivative $\mathcal{T}(\widehat{x})$ can be evaluated for all $\widehat{x} \in \Omega^M$. The information provided by $\mathcal{T}(\widehat{x})$ can be potentially used in a number of practical applications such as, for example, the shape-topological design of microtools. In particular, some numerical experiments of topology optimization of piezoelectric actuators have been presented.

Chapter 9
Asymptotic Analysis of Variational Inequalities

Contact problems with given friction are considered for plane elasticity in the framework of shape-topological optimization. The asymptotic analysis of the second kind variational inequalities in plane elasticity is performed for the purposes of shape-topological optimization. To this end, the saddle point formulation for the associated Lagrangian is introduced for the variational inequality. The non-smooth term in the energy functional is replaced by pointwise constraints for the multipliers. The one term expansion of the strain energy with respect to the small parameter which governs the size of the singular perturbation of geometrical domain is obtained. The topological derivatives of energy functional are derived in closed form adapted to the numerical methods of shape-topological optimization. In general, the topological derivative (TD) of the elastic energy is defined through a limit passage when the small parameter governing the size of the topological perturbation goes to zero. TD can be used as a steepest-descent direction in an optimization process like in any method based on the gradient of the cost functional. In this chapter, we deal with the topological asymptotic analysis in the context of contact problems with given friction. Since the problem is nonlinear, the domain decomposition technique combined with the Steklov–Poincaré pseudo-differential boundary operator is used for asymptotic analysis purposes with respect to the small parameter associated with the size of the topological perturbation. As a fundamental result, the expansion of the strain energy coincides with the expansion of the Steklov–Poincaré operator on the boundary of the truncated domain, leading to the expression for TD. Finally, the obtained TD is applied in the context of topology optimization of mechanical structures under contact condition with given friction. This chapter is based on the paper by Cinthia Gomes Lopes, Renatha Batista dos Santos, Antonio André Novotny and Jan Sokolowski [178].

A. A. Novotny et al., *Applications of the Topological Derivative Method*,
Studies in Systems, Decision and Control 188,
https://doi.org/10.1007/978-3-030-05432-8_9

9.1 Motivation

The optimum design in structural mechanics for problems governed by variational inequalities is considered in the literature using the energy functionals. Such variational problems are non-smooth, therefore one cannot expect the existence of classical shape gradients for general shape functionals depending on solutions of variational inequalities. We refer to the monograph [245] for the so-called conical shape derivatives of solutions to variational inequalities of the second kind. To this end, the solutions are given by the saddle-points of Lagrangian. The multipliers associated with the nondifferentiable terms of the elastic energy functional are subject to pointwise inequality constraints. The results obtained in [245] on shape sensitivity analysis are extended to the framework of topological sensitivity analysis. Actually, the topological derivatives of the energy functionals for contact problems with given friction with respect to the singular domain perturbations by creation of holes or inclusions are obtained. In this way, the asymptotic analysis is applied to numerical solution of optimum design for contact problems.

The complete theory of topological derivatives for linear elasticity in three spatial dimensions from the point of view of asymptotic analysis is given in [212], see also [164, 213] for further developments on polarization tensors in elasticity or piezoelasticity. The results obtained for linear elasticity cannot be directly extended to variational inequalities. The difficulty, is non-smooth nature of variational inequalities. We refer to [39] for a result obtained in the case of the Signorini problem by using the classical approach of compound asymptotic expansions under the hypothesis of strict complementarity for the unknown solution of variational inequality.

In order to circumvent this difficulty, the new method of asymptotic analysis for variational inequalities based on the domain decomposition technique combined with the compound asymptotic expansions is proposed in [242] and it is presented with all details in monograph [219]. In this way, the topological derivatives of the non-smooth energy functional can be obtained. We show also that the theoretical results on asymptotic analysis are useful for numerical solution of an important optimum design problem.

Located in Florianópolis-Brazil, the Hercílio Luz bridge shown in Fig. 9.1 is a rare and significant bridge on many different aspects. It is recognized as the longest suspension bridge in Brazil. It was also the longest spanning eyebars suspension bridge in the world when built, between 1922 and 1926. One of the most noteworthy features of Hercílio Luz bridge is that the main cables are formed by eyebars chain rather than wire cables [131].

The Hercílio Luz bridge links Florianópolis island to the continent. Because of the very aggressive environment over the ocean, it has started to suffer from a high corrosion process. In particular, some of the eyebars in the chain have collapsed according to the red line shown in Fig. 9.1. Based on safety concerns, the Hercílio Luz bridge was closed for the first time in 1982, and reopened again in 1988. After a technical report analyzing the feasibility of keeping the traffic over the bridge, presented 1990, it was completely closed in 1991. Nowadays there is an effort on

Fig. 9.1 Hercílio Luz bridge in Florianópolis, Brazil. In highlight an eyebar chain (adapted from [223])

the rehabilitation of the bridge. Another famous bridge of similar design, the Silver Bridge over the Ohio River in the U.S.A., collapsed in 1967 due to a failure of a single eyebar in the suspension chain.

In this chapter, we are interested in the redesign of an eyebar belonging to the eyebars chain of the Hercílio Luz cable bridge. The eyebars are linked through pin-joints, which are under contact condition with friction. There is a vast literature dealing with contact problems in elasticity. For the mathematical and numerical analysis of variational inequalities, see for instance the following monographs [48, 90, 114].

In order to deal with the design problem, the topology optimization of elastic structures under contact condition with given friction (stick-sleep condition) is considered. From mathematical point of view the model considered takes the form of a variational inequality of the second kind. The convenience for topological sensitivity analysis is an equivalent variational formulation as a saddle point of the Lagrangian. Such a formulation is already analyzed in [245] for the purposes of the shape sensitivity analysis. In this article new results on the existence of topological derivatives for the energy functional are derived. What is also important, the results obtained by the asymptotic analysis are used for the numerical solution of the shape-topological optimization problem. The chapter is written in such a way that it is also accessible to the engineering community. We combine the asymptotic analysis in singularly perturbed geometrical domains which belongs to pure mathematics, with the numerical methods of shape-topological optimization which belongs to applied mathematics.

Optimization of structures submitted to contact boundary conditions has received considerable attention in the last decades [49, 96, 125, 182, 246]. In particular, we are interested in the topological derivative concept [219], which is defined as the first term (correction) of the asymptotic expansion of a given shape functional with respect to a small parameter that measures the size of singular domain perturbations, such as holes, inclusions, defects, source-terms and cracks. The topological derivative can naturally be used as a steepest-descent direction in an optimization process like in any method based on the gradient of the cost functional. Therefore, this relatively new concept has applications in many different fields such as shape and topology optimization, inverse problems, imaging processing, multi-scale material design and mechanical modeling including damage, fracture evolution phenomena and control of crack propagation. See, for instance, [22, 53, 80, 211, 213, 229]. See also recent papers [113, 145, 168, 242] dealing with topological asymptotic analysis in the context of contact problems. An application of the same technique in the context of coupled electro-mechanical system can be found in [10]. These results are here extended to the case of given friction condition.

The asymptotic analysis of linear elasticity system in truncated domain has been performed in [105] by an application of the Green's function technique. In particular, the statement on a spherical hole can be found in Section 3.3, page 1766. In contrast to [105], the method developed in [242] has been designed for the purpose of asymptotic analysis in singularly perturbed domains for a class of nonlinear elasticity systems. It relies on the knowledge of the explicit solution of an auxiliary elasticity problem posed in a subdomain of simple geometry. In the ring in two spatial dimensions it is obtained by the complex Kolosov potentials [196]. By a result from the functional analysis on positive, self-adjoint operators, the expansion of the elastic energy in the ring gives rise to the expansion of the Steklov–Poincaré operator on the boundary of the topologically perturbed truncated domain, with the remainder uniformly bounded in the operator norm. In addition, the explicit solution allows us to replace the expression of the topological derivative unbounded in the energy norm by its equivalent form which is bounded in the energy norm. In this way, the truncated domain technique proposed in [105] was extended to the nonlinear contact problems in elasticity fully developed in [242]. We refer also [205] for the general case of elliptic systems and for the self-adjoint extensions of elliptic operators in punctured domains.

Therefore, following the original ideas presented by [242], in this chapter the topological derivative is extended to the context of topology optimization of elastic structures under contact condition with given friction. Since the problem is nonlinear, the domain decomposition technique combined with the Steklov–Poincaré pseudo-differential boundary operator is used in the asymptotic analysis with respect to the small parameter associated with the size of the topological perturbation. As a fundamental result, the expansion of the strain energy coincides with the expansion of the Steklov–Poincaré operator on the boundary of the truncated domain, leading to the associated topological derivative. Finally, the obtained result is used in the redesign of the eyebar from Hercílio Luz cable bridge.

9.2 Problem Formulation

Let us consider an open and bounded domain $\Omega \subset \mathbb{R}^2$ with Lipschitz boundary Γ, as shown in the sketch of Fig. 9.2. The boundary Γ consists of three mutually disjoint parts, namely, $\Gamma = \Gamma_D \cup \Gamma_N \cup \Gamma_C$. Displacements and boundary tractions

Fig. 9.2 Domain representation

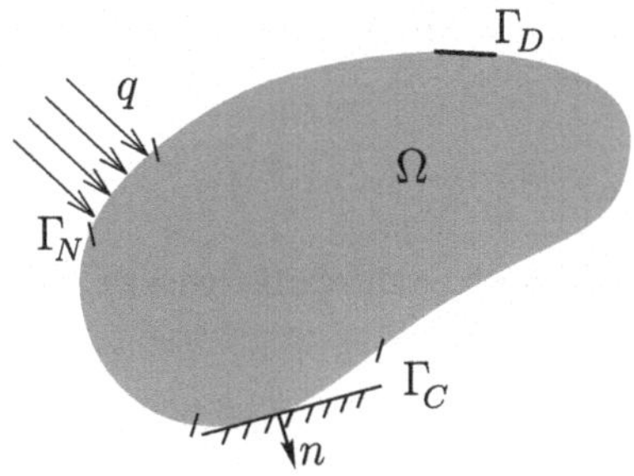

are respectively prescribed on Γ_D and Γ_N, while on Γ_C there is a possible contact condition over a rigid foundation. We assume that the normal vector on Γ_C of both elastic and rigid surfaces are collinear, allowing to set just one normal vector n on the potential contact region Γ_C. Therefore, the mechanical problem consists in finding the minimizer $u \in \mathscr{K}$ of the following functional

$$\mathscr{J}(v) := \frac{1}{2}\int_{\Omega} \sigma(v)\cdot\nabla^s v - \int_{\Gamma_N} q\cdot v + \mu_a \int_{\Gamma_C} |v\cdot\tau|, \quad \forall v \in \mathscr{K}, \tag{9.1}$$

where $\sigma(v)$ is the Cauchy stress tensor, $q \in H^{1/2}(\Gamma_N;\mathbb{R}^2)$ is a given boundary traction and μ_a is a known friction coefficient. In addition, τ denotes the tangential vector on Γ and $\mathscr{K}$ is a convex and closed cone defined as

$$\mathscr{K} := \mathscr{K}(\Omega) = \{v \in H^1(\Omega;\mathbb{R}^2) : v = 0 \text{ on } \Gamma_D \text{ and } v\cdot n \leq 0 \text{ on } \Gamma_C\}. \tag{9.2}$$

In particular, the unique minimizer $u \in \mathscr{K}$ of (9.1) is solution of the following variational inequality

$$\int_{\Omega} \sigma(u)\cdot\nabla^s(v-u) - \int_{\Gamma_N} q\cdot(v-u) + \mu_a\int_{\Gamma_C}(|v\cdot\tau| - |u\cdot\tau|) \geq 0, \quad \forall v \in \mathscr{K}. \tag{9.3}$$

From the inequality (9.3) it follows that the strong form of the elasticity equilibrium problem under contact and stick-sleep conditions is stated as [237]: Find $u : \Omega \to \mathbb{R}^2$ such that,

$$\left\{\begin{array}{rll} -\mathrm{div}(\sigma(u)) &= 0 & \text{in } \Omega, \\ \sigma(u) &= \mathbb{C}\nabla^s u & \text{in } \Omega, \\ u &= 0 & \text{on } \Gamma_D, \\ \sigma(u)n &= q & \text{on } \Gamma_N, \\ \left.\begin{array}{rl} u\cdot n &\leq 0 \\ \sigma^{nn}(u) &\leq 0 \\ \sigma^{nn}(u)(u\cdot n) &= 0 \\ \sigma^{n\tau}(u)(u\cdot\tau) + \mu_a|u\cdot\tau| &= 0 \\ -\mu_a \leq \sigma^{n\tau}(u) &\leq \mu_a \end{array}\right\} & & \text{on } \Gamma_C. \end{array}\right. \tag{9.4}$$

Some terms in the above expressions still require explanation. The strain tensor $\nabla^s u$ is given by

$$\nabla^s u := (\nabla u)^s = \frac{1}{2}(\nabla u + (\nabla u)^\top). \tag{9.5}$$

The fourth order elastic tensor $\mathbb{C} = \mathbb{C}^\top$ is written as

$$\mathbb{C} = 2\mu \mathbb{I} + \lambda(\mathrm{I} \otimes \mathrm{I}), \tag{9.6}$$

with $\mathbb{I}$ and I representing the fourth and second order identity tensors, respectively, while μ, λ are used to denote the Lamé's coefficients. In addition, the normal component $\sigma^{nn}(u)$ of the stress tensor is defined as

$$\sigma^{nn}(u) := \sigma(u)n \cdot n, \tag{9.7}$$

while the shear component on the tangential plane $\sigma^{n\tau}(u)$ is given by

$$\sigma^{n\tau}(u) := \sigma(u)n \cdot \tau. \tag{9.8}$$

Note that the contact problem we are dealing with is a simplified model where the friction coefficient is assumed to be known. The reader may refer to the papers [89, 113, 123, 125] and the book [255] for an account on similar as well as more sophisticated models.

9.3 Domain Decomposition Technique

We start by decomposing Ω into two parts, namely $\Omega = \Omega_R \cup \overline{B_R}$, where $\Omega_R := \Omega \setminus \overline{B_R}$ and B_R, with boundary Γ_R, is used to denote a ball of radius $R > 0$ centered at an arbitrary point $\widehat{x} \in \Omega$. See sketch in Fig. 9.3. Then, we consider the following linear elasticity system in B_R: Given $\psi \in H^{1/2}(\Gamma_R; \mathbb{R}^2)$, find the displacement $w : B_R \mapsto \mathbb{R}^2$, such that

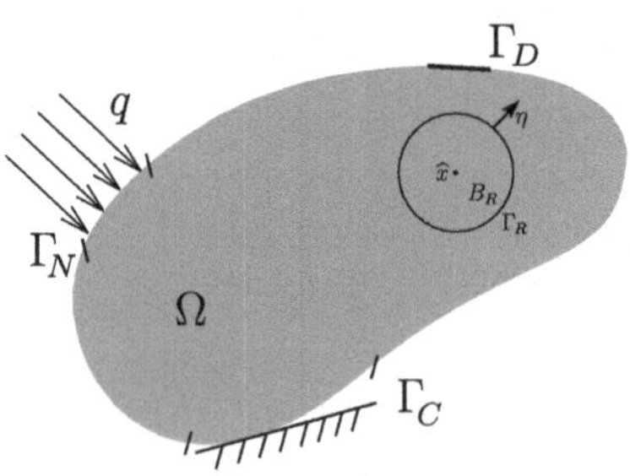

Fig. 9.3 Domain decomposition representation

$$\begin{cases} -\text{div}(\sigma(w)) = 0 & \text{in } B_R, \\ \sigma(w) = \mathbb{C}\nabla^s w & \text{in } B_R, \\ w = \psi & \text{on } \Gamma_R. \end{cases} \tag{9.9}$$

Using (9.9), we can define the Steklov–Poincaré pseudo-differential boundary operator:

$$\mathscr{A} : \psi \in H^{1/2}(\Gamma_R; \mathbb{R}^2) \mapsto \sigma(w)\eta \in H^{-1/2}(\Gamma_R; \mathbb{R}^2) \tag{9.10}$$

where η is the outward normal vector to the boundary Γ_R. Observe that by the definition of the operator $\mathscr{A}$, the solution w of (9.9) satisfies:

$$\int_{B_R} \sigma(w) \cdot \nabla^s w = \int_{\Gamma_R} \mathscr{A}(\psi) \cdot \psi. \tag{9.11}$$

That is, the energy inside B_R is equal to the energy associated with the Steklov–Poincaré operator on the boundary Γ_R. In addition, by setting $\psi = u_{|\Gamma_R}$, we have $w = u_{|B_R}$ and $u^R = u_{|\Omega_R}$. Thus,

$$\mathscr{J}(u) = \mathscr{J}^R(u^R), \tag{9.12}$$

where the functional $\mathscr{J}^R(u^R)$ is defined as

$$\mathscr{J}^R(u^R) := \frac{1}{2}\int_{\Omega_R} \sigma(u^R) \cdot \nabla^s u^R - \int_{\Gamma_N} q \cdot u^R + \mu_a \int_{\Gamma_C} |u^R \cdot \tau| + \frac{1}{2}\int_{\Gamma_R} \mathscr{A}(u^R) \cdot u^R, \tag{9.13}$$

with the minimizer $u^R \in \mathscr{K}^R := \mathscr{K}(\Omega_R)$ solution to the following variational inequality

$$\int_{\Omega_R} \sigma(u^R) \cdot \nabla^s(v - u^R) - \int_{\Gamma_N} q\cdot(v - u^R) + \mu_a \int_{\Gamma_C} (|v \cdot \tau| - |u^R \cdot \tau|) \\ + \int_{\Gamma_R} \mathscr{A}(u^R) \cdot (v - u^R) \geq 0, \quad \forall v \in \mathscr{K}^R. \tag{9.14}$$

9.4 Topological Asymptotic Analysis

We are interested on the topological derivative of the energy shape functional (9.1) with respect to the nucleation of a small inclusion. Therefore, let us consider that the domain Ω is subjected to a topological perturbation confined in a small arbitrary shaped set $\omega_\varepsilon(\widehat{x})$ of size ε and center at an arbitrary point $\widehat{x}$ of Ω, such that $\overline{\omega_\varepsilon(\widehat{x})} \subset \Omega$. In this work, the topological perturbation is characterized by the nucleation of a small circular inclusion $\omega_\varepsilon(\widehat{x}) := B_\varepsilon$ of radius $0 < \varepsilon < R$ and center at $\widehat{x} \in \Omega$, which is assumed to be far enough from the potential contact region Γ_C. This inclusion is filled with different material property from the background represented by a piecewise constant function γ_ε defined as

$$\gamma_\varepsilon = \gamma_\varepsilon(x) := \begin{cases} 1, & \text{if } x \in \Omega \setminus \overline{B_\varepsilon}, \\ \gamma, & \text{if } x \in B_\varepsilon, \end{cases} \tag{9.15}$$

with $\gamma \in \mathbb{R}^+$ used to represent the contrast on the material properties. Then, the corresponding perturbed problem consists in finding the minimizer $u_\varepsilon \in \mathscr{K}$ of the functional

$$\mathscr{J}_\varepsilon(v) := \frac{1}{2}\int_\Omega \gamma_\varepsilon \sigma(v) \cdot \nabla^s v - \int_{\Gamma_N} q \cdot v + \mu_a \int_{\Gamma_C} |v \cdot \tau|, \quad \forall v \in \mathscr{K}. \tag{9.16}$$

The element $u_\varepsilon \in \mathscr{K}$ is solution of the following perturbed variational inequality

$$\int_\Omega \gamma_\varepsilon \sigma(u_\varepsilon) \cdot \nabla^s(v - u_\varepsilon) - \int_{\Gamma_N} q \cdot (v - u_\varepsilon) + \mu_a \int_{\Gamma_C} (|v \cdot \tau| - |u_\varepsilon \cdot \tau|) \geq 0, \quad \forall v \in \mathscr{K}. \tag{9.17}$$

The associated strong system reads: Find $u_\varepsilon : \Omega \to \mathbb{R}^2$ such that,

$$\begin{cases} -\mathrm{div}(\gamma_\varepsilon \sigma(u_\varepsilon)) = 0 & \text{in } \Omega, \\ \sigma(u_\varepsilon) = \mathbb{C}\nabla^s u_\varepsilon & \text{in } \Omega, \\ u_\varepsilon = 0 & \text{on } \Gamma_D, \\ \sigma(u_\varepsilon)n = q & \text{on } \Gamma_N, \\ [\![u_\varepsilon]\!] = 0 & \text{on } \partial B_\varepsilon, \\ [\![\gamma_\varepsilon \sigma(u_\varepsilon)]\!]n = 0 & \text{on } \partial B_\varepsilon, \\ \left.\begin{array}{r} u_\varepsilon \cdot n \leq 0 \\ \sigma^{nn}(u_\varepsilon) \leq 0 \\ \sigma^{nn}(u_\varepsilon)(u_\varepsilon \cdot n) = 0 \\ \sigma^{n\tau}(u_\varepsilon)(u_\varepsilon \cdot \tau) + \mu_a |u_\varepsilon \cdot \tau| = 0 \\ -\mu_a \leq \sigma^{n\tau}(u_\varepsilon) \leq \mu_a \end{array}\right\} & \text{on } \Gamma_C. \end{cases} \tag{9.18}$$

Now, we apply the domain decomposition technique to the above topologically perturbed problem, as shown in the sketch of Fig. 9.4. This decomposition allows us to proceed with the topological asymptotic analysis in a simple geometrical domain, which is separated from the analysis of the shape function (9.16) endowed with a nondifferentiable term. Therefore, in the ball B_R we consider the following linear elasticity system associated with the perturbed domain: Given $\psi \in H^{1/2}(\Gamma_R; \mathbb{R}^2)$, find the displacement $w_\varepsilon : B_R \mapsto \mathbb{R}^2$, such that

$$\begin{cases} -\mathrm{div}(\gamma_\varepsilon \sigma(w_\varepsilon)) = 0 & \text{in } B_R, \\ \sigma(w_\varepsilon) = \mathbb{C}\nabla^s w_\varepsilon & \text{in } B_R, \\ w_\varepsilon = \psi & \text{on } \Gamma_R, \\ [\![w_\varepsilon]\!] = 0 & \text{on } \partial B_\varepsilon, \\ [\![\gamma_\varepsilon \sigma(w_\varepsilon)]\!]n = 0 & \text{on } \partial B_\varepsilon. \end{cases} \tag{9.19}$$

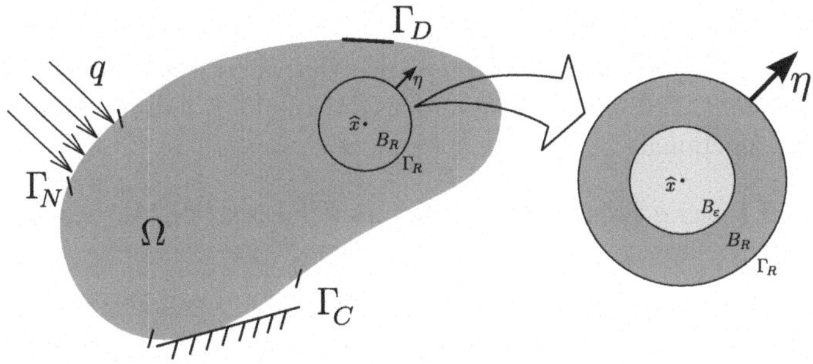

Fig. 9.4 Perturbed domain decomposition

Using problem (9.19), we define the topologically perturbed counterpart of the Steklov–Poincaré boundary operator $\mathscr{A}_\varepsilon : H^{1/2}(\Gamma_R; \mathbb{R}^2) \mapsto H^{-1/2}(\Gamma_R; \mathbb{R}^2)$ as follows

$$\mathscr{A}_\varepsilon(\psi) = \sigma(w_\varepsilon)\eta \quad \text{on} \quad \Gamma_R. \tag{9.20}$$

By the definition of the operator $\mathscr{A}_\varepsilon$, the solution w_ε of (9.19) satisfies:

$$\int_{B_R} \gamma_\varepsilon \sigma(w_\varepsilon) \cdot \nabla^s w_\varepsilon = \int_{\Gamma_R} \mathscr{A}_\varepsilon(\psi) \cdot \psi. \tag{9.21}$$

That is, the energy inside the ball $B_R \supset B_\varepsilon$ is equal to the energy associated with the Steklov–Poincaré operator $\mathscr{A}_\varepsilon$ on Γ_R. We observe that by setting $\psi = u_{\varepsilon|_{\Gamma_R}}$ we have $w_\varepsilon = u_{\varepsilon|_{B_R}}$ and $u_\varepsilon^R = u_{\varepsilon|_{\Omega_R}}$, which implies the equality

$$\mathscr{J}_\varepsilon(u_\varepsilon) = \mathscr{J}_\varepsilon^R(u_\varepsilon^R), \tag{9.22}$$

where the functional $\mathscr{J}_\varepsilon^R(u_\varepsilon^R)$ is written as

$$\mathscr{J}_\varepsilon^R(u_\varepsilon^R) := \frac{1}{2}\int_{\Omega_R} \sigma(u_\varepsilon^R) \cdot \nabla^s u_\varepsilon^R - \int_{\Gamma_N} q \cdot u_\varepsilon^R + \mu_a \int_{\Gamma_C} |u_\varepsilon^R \cdot \tau| + \frac{1}{2}\int_{\Gamma_R} \mathscr{A}_\varepsilon(u_\varepsilon^R) \cdot u_\varepsilon^R, \tag{9.23}$$

with the minimizer $u_\varepsilon^R \in \mathscr{K}^R$ given by the unique solution of the variational inequality:

$$\int_{\Omega_R} \sigma(u_\varepsilon^R) \cdot \nabla^s (v - u_\varepsilon^R) - \int_{\Gamma_N} q \cdot (v - u_\varepsilon^R) + \mu_a \int_{\Gamma_C} (|v \cdot \tau| - |u_\varepsilon^R \cdot \tau|) + \int_{\Gamma_R} \mathscr{A}_\varepsilon(u_\varepsilon^R) \cdot (v - u_\varepsilon^R) \geq 0, \quad \forall v \in \mathscr{K}^R. \tag{9.24}$$

Before evaluate the topological derivative, we present two important results. The first one ensures the existence of the topological derivative associated with the problem under analysis. The second result proves the topological differentiability of the energy shape functional.

Proposition 9.1 *Let u^R and u_ε^R be solution to* (9.17) *and* (9.24), *respectively, and assume that*

$$\mathscr{A}_\varepsilon = \mathscr{A} - \varepsilon^2 \mathscr{B} + \mathscr{R}_\varepsilon, \tag{9.25}$$

in the operator norm $\mathscr{L}(H^{1/2}(\Gamma_R;\mathbb{R}^2); H^{-1/2}(\Gamma_R;\mathbb{R}^2))$, with $\mathscr{B}$ used to denote a bounded linear operator and

$$\|\mathscr{R}_\varepsilon\|_{\mathscr{L}(H^{1/2}(\Gamma_R;\mathbb{R}^2);H^{-1/2}(\Gamma_R;\mathbb{R}^2))} = o(\varepsilon^2). \tag{9.26}$$

Then, the following estimate holds true:

$$\|u_\varepsilon^R - u^R\|_{H^1(\Omega^R;\mathbb{R}^2)} \le C\varepsilon^2. \tag{9.27}$$

Proof By taking $v = u_\varepsilon^R$ in (9.17) and $v = u^R$ in (9.24), we obtain:

$$\int_{\Omega_R} \sigma(u^R)\cdot\nabla^s(u_\varepsilon^R - u^R) - \int_{\Gamma_N} q\cdot(u_\varepsilon^R - u^R) + \\ \mu_a \int_{\Gamma_C} (|u_\varepsilon^R\cdot\tau| - |u^R\cdot\tau|) + \int_{\Gamma_R} \mathscr{A}(u^R)\cdot(u_\varepsilon^R - u^R) \ge 0 \tag{9.28}$$

and

$$\int_{\Omega_R} \sigma(u_\varepsilon^R)\cdot\nabla^s(u^R - u_\varepsilon^R) - \int_{\Gamma_N} q\cdot(u^R - u_\varepsilon^R) + \\ \mu_a \int_{\Gamma_C} (|u^R\cdot\tau| - |u_\varepsilon^R\cdot\tau|) + \int_{\Gamma_R} \mathscr{A}_\varepsilon(u_\varepsilon^R)\cdot(u^R - u_\varepsilon^R) \ge 0. \tag{9.29}$$

Then, after adding (9.28) and (9.29), we write

$$\int_{\Omega_R} \sigma(u_\varepsilon^R - u^R)\cdot\nabla^s(u_\varepsilon^R - u^R) + \int_{\Gamma_R} [\mathscr{A}_\varepsilon(u_\varepsilon^R) - \mathscr{A}(u^R)]\cdot(u_\varepsilon^R - u^R) \le 0. \tag{9.30}$$

Now, assuming (9.25), we have that

$$\int_{\Omega_R} \sigma(u_\varepsilon^R - u^R) \cdot \nabla^s(u_\varepsilon^R - u^R) + \int_{\Gamma_R} \mathscr{A}(u_\varepsilon^R - u^R) \cdot (u_\varepsilon^R - u^R) \leq \varepsilon^2 \int_{\Gamma_R} \mathscr{B}(u_\varepsilon^R) \cdot (u_\varepsilon^R - u^R) + o(\varepsilon^2), \tag{9.31}$$

and by the coercivity of the bilinear form on the left-hand side of (9.31) it follows

$$C_0 \|u_\varepsilon^R - u^R\|^2_{H^1(\Omega_R;\mathbb{R}^2)} \leq \varepsilon^2 \int_{\Gamma_R} \mathscr{B}(u_\varepsilon^R) \cdot (u_\varepsilon^R - u^R) + o(\varepsilon^2). \tag{9.32}$$

And also,

$$\begin{aligned} \|u_\varepsilon^R - u^R\|^2_{H^1(\Omega_R;\mathbb{R}^2)} &\leq C_1 \varepsilon^2 \|\mathscr{B}(u_\varepsilon^R)\|_{H^{-1/2}(\Gamma_R;\mathbb{R}^2)} \|u_\varepsilon^R - u^R\|_{H^{1/2}(\Gamma_R;\mathbb{R}^2)} \\ &\leq C_2 \varepsilon^2 \|u_\varepsilon^R - u^R\|_{H^1(\Omega_R;\mathbb{R}^2)}. \end{aligned} \tag{9.33}$$

Finally, we obtain (9.27) with $C = C_2/C_0$ independent of the small parameter ε.

Lemma 9.1 *The perturbed energy shape functional $\mathscr{J}_\varepsilon^R(u_\varepsilon^R)$ in (9.23) is differentiable with respect to $\varepsilon \to 0$. In particular, it admits the asymptotic expansion*

$$\mathscr{J}_\varepsilon^R(u_\varepsilon^R) = \mathscr{J}^R(u^R) - \frac{\varepsilon^2}{2} \langle \mathscr{B}(u^R), u^R \rangle_{\Gamma_R} + o(\varepsilon^2), \tag{9.34}$$

where $\langle \phi, \varphi \rangle_{\Gamma_R}$ is used to denote the inner product on Γ_R, that is

$$\langle \phi, \varphi \rangle_{\Gamma_R} = \int_{\Gamma_R} \phi \cdot \varphi. \tag{9.35}$$

Proof By taking into account that $u_\varepsilon^R \in \mathscr{K}^R$ is the minimizer of (9.23) and $u^R \in \mathscr{K}^R$ is the minimizer of (9.13), the following inequalities hold true

$$\mathscr{J}_\varepsilon^R(u_\varepsilon^R) - \mathscr{J}^R(u_\varepsilon^R) \leq \mathscr{J}_\varepsilon^R(u_\varepsilon^R) - \mathscr{J}^R(u^R) \leq \mathscr{J}_\varepsilon^R(u^R) - \mathscr{J}^R(u^R). \tag{9.36}$$

Using the definitions of $\mathscr{J}_\varepsilon^R$ and $\mathscr{J}^R$, considering the expansion (9.25) and after organizing all the terms, we have

$$\frac{\mathscr{J}_\varepsilon^R(u^R) - \mathscr{J}^R(u^R)}{\varepsilon^2} = -\frac{1}{2} \langle \mathscr{B}(u^R), u^R \rangle_{\Gamma_R} + \langle \mathscr{R}_\varepsilon(u^R), u^R \rangle_{\Gamma_R}. \tag{9.37}$$

Thus, it follows that

$$\lim_{\varepsilon \to 0} \left(\frac{\mathscr{J}\varepsilon^R(u^R) - \mathscr{J}^R(u^R)}{\varepsilon^2} \right) = -\frac{1}{2} \langle \mathscr{B}(u^R), u^R \rangle_{\Gamma_R}. \tag{9.38}$$

Similarly, we write

$$\frac{\mathscr{J}_\varepsilon^R(u_\varepsilon^R) - \mathscr{J}^R(u_\varepsilon^R)}{\varepsilon^2} = -\frac{1}{2}\langle \mathscr{B}(u_\varepsilon^R), u_\varepsilon^R \rangle_{\Gamma_R} + \langle \mathscr{R}_\varepsilon(u_\varepsilon^R), u_\varepsilon^R \rangle_{\Gamma_R}. \tag{9.39}$$

Now, taking into account the strong convergence of the minimizers in the energy space, Proposition 9.1, we obtain

$$\lim_{\varepsilon\to 0}\left(\frac{\mathscr{J}_\varepsilon^R(u_\varepsilon^R) - \mathscr{J}^R(u_\varepsilon^R)}{\varepsilon^2}\right) = -\frac{1}{2}\langle \mathscr{B}(u^R), u^R \rangle_{\Gamma_R}. \tag{9.40}$$

From the above limits and (9.36) we conclude that

$$\lim_{\varepsilon\to 0}\left(\frac{\mathscr{J}_\varepsilon^R(u_\varepsilon^R) - \mathscr{J}^R(u^R)}{\varepsilon^2}\right) = -\frac{1}{2}\langle \mathscr{B}(u^R), u^R \rangle_{\Gamma_R}. \tag{9.41}$$

Therefore, we can write (9.34).

9.5 Topological Derivative Formula

In this section, the topological derivative associated with the problem under analysis is obtained in its closed form. Before proceeding, let us state the following important result, whose proof can be found in [219].

Lemma 9.2 *The energy inside B_R admits the asymptotic expansion:*

$$\int_{B_R} \sigma_\varepsilon(w_\varepsilon)\cdot\nabla^s(w_\varepsilon) = \int_{B_R} \sigma(w)\cdot\nabla^s(w) - \varepsilon^2\mathbb{P}_\gamma\sigma(w)\cdot\nabla^s w + o(\varepsilon^2), \tag{9.42}$$

where the polarization tensor $\mathbb{P}_\gamma$ is given by the following fourth order isotropic tensor

$$\mathbb{P}_\gamma = \frac{\pi(1-\gamma)}{1+\gamma a_2}\left((1+a_2)\mathbb{I} + \frac{1}{2}(a_1 - a_2)\frac{1-\gamma}{1+\gamma a_1}\mathrm{I}\otimes\mathrm{I}\right), \tag{9.43}$$

with $0<\gamma<\infty$ and the parameters a_1, a_2 given by

$$a_1 = \frac{\lambda+\mu}{\mu} \quad \textit{and} \quad a_2 = \frac{\lambda+3\mu}{\lambda+\mu}. \tag{9.44}$$

Now, note that by Proposition 9.1 we write

$$\langle \mathscr{A}_\varepsilon(\phi), \varphi\rangle = \langle \mathscr{A}(\phi), \varphi\rangle - \varepsilon^2\langle \mathscr{B}(\phi), \varphi\rangle + o(\varepsilon^2), \quad \forall \phi,\ \varphi. \tag{9.45}$$

Then, from Lemma 9.2 and expansion (9.45) we conclude that the expansion of the strain energy in B_R coincides with the expansion of the Steklov–Poincaré operator on the boundary Γ_R. Therefore, it follows that

$$\int_{\Gamma_R} \mathscr{B}(\phi) \cdot \psi = \mathbb{P}_\gamma \sigma(\phi) \cdot \nabla^s \varphi, \quad \forall \phi, \varphi. \tag{9.46}$$

Thus, by Lemma 9.1 together with Eq. (9.46) we have

$$\mathscr{J}_\varepsilon(u_\varepsilon) - \mathscr{J}(u) = -\frac{1}{2}\varepsilon^2 \mathbb{P}_\gamma \sigma(u) \cdot \nabla^s u + o(\varepsilon^2), \tag{9.47}$$

where we have also considered the equalities (9.12) and (9.22). Finally, by choosing $f(\varepsilon) = \varepsilon^2$, we have the main result of the chapter, namely:

Theorem 9.1 *The topological derivative of the shape functional* $\mathscr{J}(u)$*, defined in* (9.1)*, is given by*

$$\mathscr{T}(x) = -\frac{1}{2}\mathbb{P}_\gamma \sigma(u(x)) \cdot \nabla^s u(x), \quad \forall x \in \Omega. \tag{9.48}$$

9.6 Numerical Application

In this section the obtained topological derivative (9.48) is applied in the context of topology optimization of structures under contact condition with given friction. The idea is to redesign an eyebar belonging to the eyebars chain of the Hercílio Luz cable bridge, as presented in Fig. 9.1.

Therefore, let us consider a hold-all domain $\mathscr{D} \subset \mathbb{R}^2$ such that $\Omega \subset \mathscr{D}$. The topology optimization problem we are dealing with consists in minimizing the total potential energy for a given amount of material, that is:

$$\begin{cases} \underset{\Omega \subset \mathscr{D}}{\text{Minimize}} \ \mathscr{J}(u) \\ \text{subject to} \ \ |\Omega| \leq M, \end{cases} \tag{9.49}$$

where $|\Omega|$ is the Lebesgue measure of Ω and M is the desired volume at the end of the optimization process. To deal with the volume constraint we use the Linear Penalization Method. Thus, problem (9.49) is rewritten as following:

$$\underset{\Omega \subset \mathscr{D}}{\text{Minimize}} \ \mathscr{F}_\Omega(u) := \mathscr{J}(u) + \beta|\Omega|, \tag{9.50}$$

where $\beta = \beta^\star / V_0$ is a penalization parameter with $\beta^\star > 0$, and V_0 denotes the initial volume of the structure. The optimization problem (9.50) is solved by using the topology optimization algorithm proposed in [25], which relies on the topological derivative concept and a level-set domain representation method. In particular, the topological derivative is used as a feasible descent direction to minimize the cost functional (9.50). The reader may refer to [10, 25, 177] for more details and applications of this algorithm. See also Algorithm 1 presented in Chap. 6.

The numerical implementation was performed by using FEniCS software [175]. Once we are interested in redesigning the eyebar shown in Fig. 9.1, we consider as original design the domain shown in Fig. 9.5a. On the other hand, the initial guess is given by a rectangle of dimensions $400 \times 1651\,\text{mm}^2$ with a semicircular hole of radius r_1, as shown in Fig. 9.5b. See also [223]. We consider vertical symmetry conditions, represented by dashed lines. The potential contact region is given by the boundary of the semicircle of radius r_1. The structure is submitted to a distributed load q as shown in the sketch of Fig. 9.5, whose resultant is denoted by $Q = 24000\,\text{N}$. In addition, the eyebar is made with steel, whose Young's modulus E, Poisson ratio ν, friction coefficient μ_a and stress limit of the material $\overline{\sigma}$ are given in Table 9.1, together with additional geometrical properties. Finally, we set the volume penalization parameter $\beta^\star = 4 \times 10^4$.

The obtained result is presented in Fig. 9.6. The normalized von Mises stresses distribution are shown in Fig. 9.7. Note that in both cases the maximal value is bounded by one. However, the original design has approximately 46.08% (orange area) while the obtained optimized eyebar has approximately 38.62% of volume

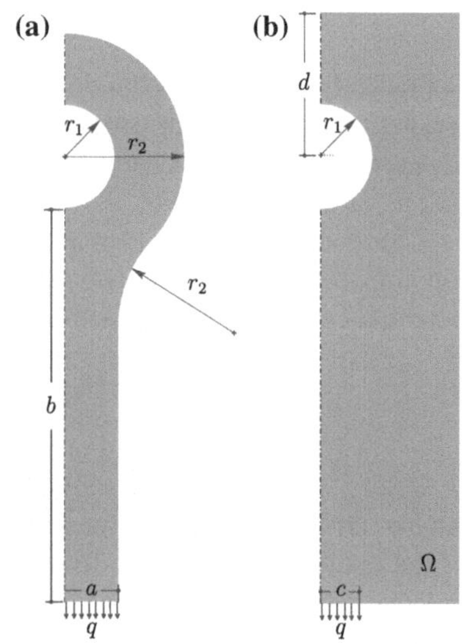

Fig. 9.5 Original design (**a**) and initial guess (**b**)

Table 9.1 Material properties and geometrical parameters

ν	μ_a	E	$\overline{\sigma}$	Q	a	b	c	d	r_1	r_2
0.3	0.2	210×10^3 MPa	440 MPa	24000 N	152 mm	850 mm	127 mm	451 mm	146 mm	350 mm

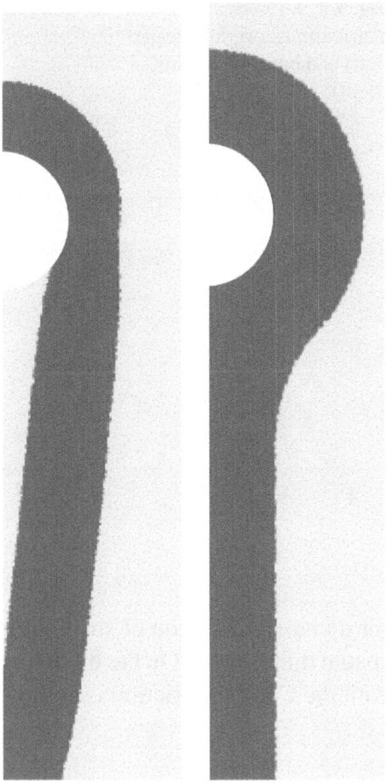

Fig. 9.6 Original design (left) and obtained result (right)

fraction, which corresponds to more than 16% of volume reduction without violate the stress limit of the material.

9.7 Final Remarks

In this chapter, the topological derivative concept has been applied in the context of contact problems in elasticity with given friction. Since the problem is nonlinear, the domain decomposition technique together with the Steklov–Poincaré pseudo-differential boundary operator were used in the topological asymptotic analysis of the energy shape functional with respect to the nucleation of a small circular inclusion. From such an analysis, the associated topological derivative has been derived in its closed form. The obtained result has been applied in a case study concerning the redesign of an eyebar belonging to the eyebars chain of the Hercílio Luz cable bridge. As a result, the obtained optimal design is much more efficient from the mechanical point of view in comparison with the original one, since its volume has been reduced about 16% while the maximal von Mises stress does not exceed the stress limit of the material. The proposed method is general and it can be applied

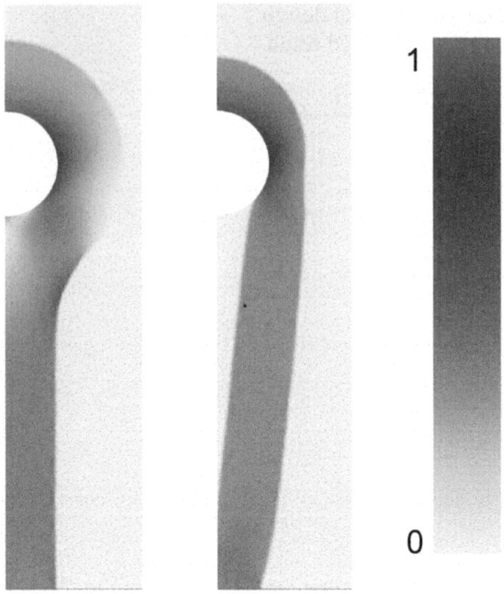

Fig. 9.7 Stresses distributions: original design (left) and optimized result (right)

for numerical solution of shape-topology optimization of contact problems in three spatial dimensions. On the other hand, the topological sensitivity analysis of contact with the Coulomb friction is still an unsolved and difficult problem.

Chapter 10
A Newton-Type Method and Applications

The framework of asymptotic analysis in singularly perturbed geometrical domains can be employed to produce two-term asymptotic expansions for a class of shape functionals. In Chap. 6 one-term expansions of functionals are required for algorithms of shape-topological optimization. Such an approach corresponds to the simple gradient method in shape optimization. The Newton method of shape optimization can be replaced, for shape-topology optimization, by two-term expansions of shape functionals. Thus, the resulting approximations are more precise and the associated numerical methods are much more complex compared to one-term expansion topological derivative algorithms. In particular, numerical algorithms associated with first order topological derivatives of shape functionals have been presented in Chap. 6, together with an account of their applications currently found in the literature, with emphasis on shape and topology optimization. In this chapter second order topological derivatives are introduced. Second order algorithms of shape-topological optimization are used for numerical solution of representative examples of inverse reconstruction problems. The main feature of these algorithms is that the method is non-iterative and thus very robust with respect to noisy data as well as independent of initial guesses.

10.1 Preliminaries

Topological sensitivity analysis leads to the asymptotic expansion of a given shape functional with respect to a small parameter that measures the size of singular domain perturbations, such as holes, inclusions, source-terms and cracks. This relatively new concept has applications in many different fields such as shape and topology optimization, inverse problems, image processing, multi-scale material design and mechanical modeling including damage and fracture evolution phenomena.

A. A. Novotny et al., *Applications of the Topological Derivative Method*,
Studies in Systems, Decision and Control 188,
https://doi.org/10.1007/978-3-030-05432-8_10

Numerical algorithms associated with first order topological derivatives have been presented in Chap. 6, together with an account of their applications currently found in the literature, with emphasis on shape and topology optimization.

In this chapter the concept of second order topological derivative is introduced [81], which is used to solve a class of inverse reconstruction problems formulated as overdetermined boundary value problems. The general idea is to rewrite them as topology optimization problems. In particular, a shape functional measuring the misfit between boundary measurements and the solution obtained from the model is minimized with respect to a set of ball-shaped anomalies by using the concept of topological derivative. This means that the objective functional is expanded and then truncated to the second order, leading to a quadratic and strictly convex form with respect to the parameters under consideration. Thus, a trivial optimization step leads to a non-iterative second order reconstruction algorithm. As a result, the reconstruction process becomes very robust with respect to noisy data and independent of any initial guess. These results are used to solve a wide class of inverse reconstruction problems.

The chapter is organized as follows. The notion of second order topological derivative is introduced in Sect. 10.2. The resulting second order method is presented in Sect. 10.3, together with the associated algorithm. In Sect. 10.4 a class of inverse reconstruction problems formulated as overdetermined boundary value problems is introduced. In particular, Sect. 10.4.1 deals with the inverse gravimetry problem. A pointwise source reconstruction problem is introduced in Sect. 10.4.2. Section 10.4.3 is dedicated to an obstacle reconstruction problem. In Sects. 10.4.4 and 10.4.5 the inverse electromagnetic casting and the electrical impedance tomography problems are discussed, respectively. Finally, the chapter ends in Sect. 10.5 with a discussion concerning perspectives of future development, together with a list of open problems.

10.2 Higher Order Topological Derivative

Let us consider an open bounded domain $\Omega \subset \mathbb{R}^d, d \geq 2$, with Lipschitz continuous boundary $\partial\Omega$. The domain Ω is subjected to a perturbation confined to a small, arbitrarily shaped set $\omega_\varepsilon(\widehat{x})$ of size ε and center at an arbitrary point $\widehat{x}$ of Ω, such that $\overline{\omega_\varepsilon(\widehat{x})} \subset \Omega$. We introduce a characteristic function $x \mapsto \chi(x), x \in \Omega$, associated to the unperturbed domain, namely $\chi = \mathbb{1}_\Omega$. Then, we define a characteristic function associated to the topologically perturbed domain of the form $x \mapsto \chi_\varepsilon(\widehat{x}; x), x \in \Omega$. In the case of a hole, for example, $\chi_\varepsilon(\widehat{x}) = \mathbb{1}_\Omega - \mathbb{1}_{\overline{\omega_\varepsilon(\widehat{x})}}$ and the perturbed domain is given by $\Omega_\varepsilon(\widehat{x}) = \Omega \setminus \overline{\omega_\varepsilon(\widehat{x})}$. Then, we assume that a given shape functional $\psi(\chi_\varepsilon(\widehat{x}))$, associated to the topologically perturbed domain, admits the following topological asymptotic expansion [219]:

$$\psi(\chi_\varepsilon(\widehat{x})) = \psi(\chi) + f_1(\varepsilon)\mathcal{T}(\widehat{x}) + f_2(\varepsilon)\mathcal{T}^2(\widehat{x}) + o(f_2(\varepsilon)) \,, \tag{10.1}$$

where $\psi(\chi)$ is the shape functional associated to the unperturbed domain, $f_1(\varepsilon)$ and $f_2(\varepsilon)$ are positive functions and $o(f_2(\varepsilon))$ is the remainder, such that $f_1(\varepsilon) \to 0$, $f_2(\varepsilon)/f_1(\varepsilon) \to 0$ and $o(f_2(\varepsilon))/f_2(\varepsilon) \to 0$ as $\varepsilon \to 0$. The functions $\widehat{x} \mapsto \mathcal{T}(\widehat{x})$ and $\widehat{x} \mapsto \mathcal{T}^2(\widehat{x})$ are called the first and second order topological derivatives of ψ at $\widehat{x}$. The terms $f_1(\varepsilon)\mathcal{T}(\widehat{x})$ and $f_2(\varepsilon)\mathcal{T}^2(\widehat{x})$ represent, respectively, first and second order corrections to $\psi(\chi)$ in order to approximate $\psi(\chi_\varepsilon(\widehat{x}))$. Therefore, the first order topological derivative $\mathcal{T}(\widehat{x})$ can naturally be used as a steepest-descent direction in an optimization process, just as in any method based on the gradient of the cost functional, leading to a family of first order topology optimization algorithms. On the other hand, the second order topological derivative $\mathcal{T}^2(\widehat{x})$ leads to second order topology optimization algorithms, which are non-iterative and thus very robust with respect to noisy data, for instance. Since all quantities on the right-hand side of (10.1) are defined in the original (unperturbed) domain Ω, the resulting algorithms derived from $\mathcal{T}(\widehat{x})$ and $\mathcal{T}^2(\widehat{x})$ are independent of initial guesses.

The form of the topological asymptotic expansion (10.1) depends on many features of the problem under consideration, including the spatial dimension, differential operator, nature of the topological perturbations and their boundary/transmission conditions, etc. To fix ideas, let us consider the elasticity problem in three spatial dimensions, whose topological perturbation is given by the nucleation of an arbitrarily shaped inclusion with material properties different from the background. According to [56], in this case the expansion (10.1) takes the form

$$\psi(\chi_\varepsilon(\widehat{x})) = \psi(\chi) + \varepsilon^3\mathcal{T}(\widehat{x}) + \varepsilon^4\mathcal{T}^2(\widehat{x}) + \varepsilon^5\mathcal{T}^3(\widehat{x}) + \varepsilon^6\mathcal{T}^4(\widehat{x}) + o(\varepsilon^6)\,. \tag{10.2}$$

For spherical or ellipsoidal shaped inclusions, the $O(\varepsilon^4)$ term vanishes. See also [130] for an equivalent expansion in the scalar case. In two spatial dimensions, topological asymptotic analysis may become more involved. In fact, depending on the problem under consideration, the logarithm of ε may appear. See for instance Sect. 10.4.3. Let us remark however that this is a rich and fascinating field of research with a wide range of relevant applications and many unsolved theoretical questions. See Sect. 10.5 for an account of some open problems.

10.3 Second Order Reconstruction Algorithm

A well known phenomenon concerning high order topological asymptotic expansions concerns interaction between several topological perturbations. Therefore, let us assume that the domain Ω is perturbed by the nucleation of N ball-shaped anomalies $B_{\varepsilon_i}(x_i)$ of radii ε_i and centers at $x_i \in \Omega$, with $i = 1, \ldots, N$. See sketch in Fig. 10.1. We introduce the notations $\xi = (x_1, \ldots, x_N)$ and $\varepsilon = (\varepsilon_1, \ldots, \varepsilon_N)$. To fix ideas, we restrict ourselves to the case where the expansion (10.1) takes the following quadratic form with respect to α:

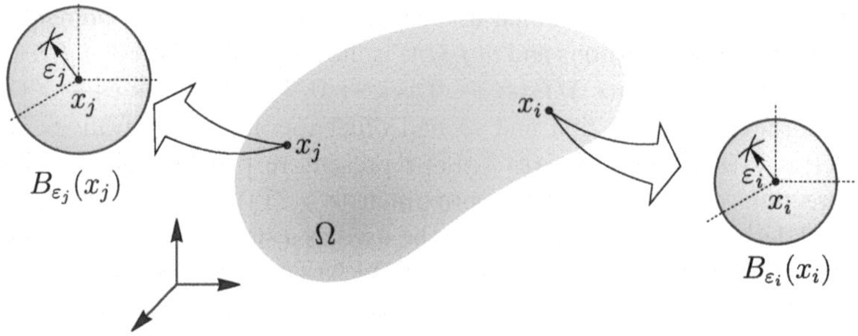

Fig. 10.1 Nucleation of a set of ball-shaped inclusions $B_{\varepsilon_i}(x_i)$, $i = 1, \ldots, N$

$$\psi(\chi_\varepsilon(\xi)) = \psi(\chi) - \alpha \cdot d(\xi) + \frac{1}{2} H(\xi)\alpha \cdot \alpha + \mathscr{E}(\varepsilon) , \tag{10.3}$$

where $\alpha = (\alpha_1, \ldots, \alpha_N)$, with $\alpha_i = f_1(\varepsilon_i)$, and $\mathscr{E}(\varepsilon)$ is a remainder. The vector $d \in \mathbb{R}^N$ and the Hessian matrix $H \in \mathbb{R}^N \times \mathbb{R}^N$ represent the first and second order topological derivatives, respectively. Their entries are d_i and H_{ij}. In addition, the expression on the right-hand side of (10.3) depends on the number N of anomalies, their size α and locations ξ. We note however that this kind of expansion strongly depends on the problem under consideration. In particular, the expansion (10.3) holds true for EIT in two spatial dimensions [101] and for gravimetry in general [68]. In contrast, it is not valid for EIT in three spatial dimensions [130] or in the case of the inverse potential problem proposed in [137, pp. 126, Problem 4.2]. Actually, in [99] the topological asymptotic expansion associated with a modified Helmholtz equation has been rigorously derived, leading to an intermediate term of the form $O(|\alpha|^2 \log \alpha)$. From (10.3), we can define the following quantity:

$$\delta J(\alpha, \xi, N) := -\alpha \cdot d(\xi) + \frac{1}{2} H(\xi)\alpha \cdot \alpha , \tag{10.4}$$

Let us assume that the Hessian matrix $H(\xi)$ is positive definite (this condition can be easily verified for each problem under consideration by following the same steps as in [68, Proposition 4], for instance), so that the minimization of the function $\delta J(\alpha, \xi, N)$ with respect to α yields

$$\langle D_\alpha \delta J, \beta \rangle = (H(\xi)\alpha - d(\xi)) \cdot \beta = 0, \quad \forall \beta , \tag{10.5}$$

which leads to the linear system

$$H(\xi)\alpha = d(\xi) . \tag{10.6}$$

The quantity α solving (10.6) becomes a function of the locations ξ, $\alpha = \alpha(\xi)$. Let us now replace the solution of (10.6) into $\delta J(\alpha, \xi, N)$ defined by (10.4). Therefore, the

Algorithm 2: Second Order Algorithm.

input : d, H, M, N;
output: the optimal solution $S^\star, \alpha^\star, \xi^\star$;

Initialization: $S^\star \leftarrow \infty$; $\alpha^\star \leftarrow 0$; $\xi^\star \leftarrow 0$;
for $i_1 \leftarrow 1$ **to** M **do**
  for $i_2 \leftarrow i_1 + 1$ **to** M **do**
    ⋮
    for $i_N \leftarrow i_{N-1} + 1$ **to** M **do**

$$d \leftarrow \begin{bmatrix} f(i_1) \\ f(i_2) \\ \vdots \\ f(i_N) \end{bmatrix}; \quad H \leftarrow \begin{bmatrix} A(i_1, i_1) & A(i_1, i_2) & \cdots & A(i_1, i_N) \\ A(i_2, i_1) & A(i_2, i_2) & \cdots & A(i_2, i_N) \\ \vdots & \vdots & \ddots & \vdots \\ A(i_N, i_1) & A(i_N, i_2) & \cdots & A(i_N, i_N) \end{bmatrix};$$

$\mathscr{I} \leftarrow (i_1, i_2, \ldots, i_N)$; $\xi \leftarrow \Pi(\mathscr{I})$; $\alpha \leftarrow H^{-1}d$; $S \leftarrow -\frac{1}{2}d \cdot \alpha$;
      if $S < S^\star$ **then**
        $\xi^\star \leftarrow \xi$; $\alpha^\star \leftarrow \alpha$; $S^\star \leftarrow S$;
      end if
    end for
  end for
end for
return $S^\star, \alpha^\star, \xi^\star$;

optimal locations $\xi^\star$ can be obtained from a combinatorial search over the domain Ω, as solutions to the following minimization problem:

$$\xi^\star = \underset{\xi \in X}{\operatorname{argmin}} \left\{ \delta J(\alpha(\xi), \xi, N) = -\frac{1}{2}\alpha(\xi) \cdot d(\xi) \right\}, \tag{10.7}$$

where X is the set of admissible locations of anomalies. Finally, the optimal intensities are given by $\alpha^\star = \alpha(\xi^\star)$. In summary, our method is able to find optimal sizes $\alpha^\star$ of the hidden anomalies and their locations $\xi^\star$ for a given number N of trial balls.

To summarize, we now introduce the resulting second order topology optimization algorithm. It describes the process of obtaining the optimal parameters $\alpha^\star$ and $\xi^\star$ from the computational point of view. The input of the algorithm is listed below:

- the number N of anomalies;
- the M points at which the systems (10.6) are solved;
- the vector d and the matrix H, whose entries are given by $f(i) := d_i$ and $A(i, j) := H_{ij}$, respectively.

The algorithm returns optimal sizes $\alpha^\star$ and locations $\xi^\star$. The above procedure written in pseudo-code format is shown in Algorithm 2. In the algorithm, Π maps the vector of nodal indices $\mathscr{I} = (i_1, i_2, \ldots, i_N)$ to the corresponding vector of nodal coordinates ξ. For further applications of this algorithm we refer to [67, 68, 101, 181, 225], for instance.

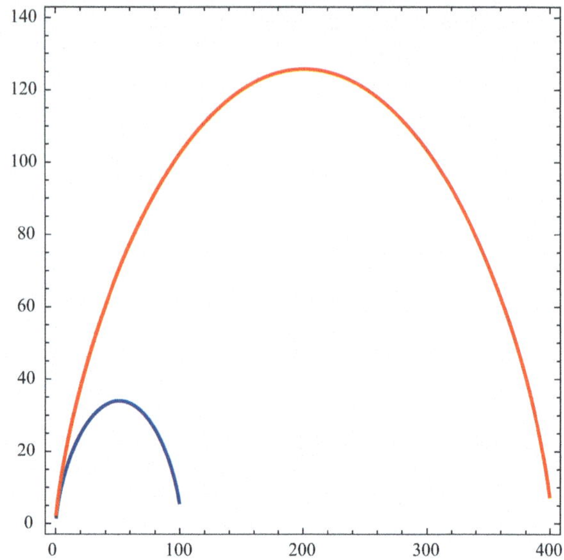

Fig. 10.2 Complexity order of Algorithm 2: $N \times \log_{10}(\mathscr{C}(M, N))$ for $M = 100$ (blue) and $M = 400$ (red)

As can be noted in Algorithm 2, the optimal solution $(\xi^\star, \alpha^\star)$ is obtained through a combinatorial and exhaustive search over the M points. Therefore, the complexity $\mathscr{C}(M, N)$ of the algorithm can be evaluated by the formula

$$\mathscr{C}(M, N) = \binom{M}{N} N^3 = \frac{M!}{N!(M-N)!} N^3. \tag{10.8}$$

In Fig. 10.2 the graphs of $N \times \log_{10}(\mathscr{C}(M, N))$ for $M = 100$ and $M = 400$ are presented in blue and red, respectively.

Since Algorithm 2 is the bottleneck of the proposed second order topology optimization method, we refer to [181] for more sophisticated approaches based on meta-heuristic and multi-grid versions of Algorithm 2. In addition, approximation of the solution by a finite number of balls can be seen as a limitation of our approach. However, the reconstruction obtained may serve as an initial guess for other well-established and more computationally sophisticated iterative methods [62, 127, 138, 163, 249].

10.4 Inverse Reconstruction Problems

In this section the topological derivative concept is applied in the context of a class of inverse problems in imaging. A wide class of inverse problems can be reformulated as overdetermined boundary value problems. This difficulty can be overcome by rewriting the inverse problem in the form of an optimization problem. The basic idea is to minimize an objective functional measuring the misfit between a given data and a numerical solution with respect to the parameters under consideration. In particular, let us consider a geometrical domain Ω with boundary $\Gamma = \partial\Omega$. A boundary value problem is defined in Ω, whose solution is denoted by u^*. We assume that the response of the system on the boundary Γ can be observed. For example, given a Dirichlet data U on Γ, the associated Dirichlet-to-Neumann map for a second order elliptic equation is defined as follows [65]:

$$\Lambda_{\omega^*} \, : \, u^* = U \mapsto Q := \partial_n u^* \quad \text{on} \quad \Gamma.$$

where ω^* is an unknown set of anomalies embedded in Ω and n is the exterior unit normal vector on Γ. Therefore, given the pair (U, Q) we want to reconstruct the set $\omega^* \subset \Omega$. The mathematical model of the system furnishes the mapping $\omega \mapsto \Lambda_\omega$ for a family of anomalies ω. Thus, taking U we can generate the output $\Lambda_\omega(U)$ and compare it with the given function $Q = \Lambda_{\omega^*}(U)$. Hence, using the mathematical model we can consider the associated optimization problem based on minimizing the distance between the observation (U, Q) and the model response $(U, \Lambda_\omega(U))$ over the family of admissible anomalies ω. This is a numerical method which uses the shape and topological derivatives of the specific shape functional defined for the inverse problem [30, 67, 75, 76, 119, 127, 130, 140, 159, 183]. In particular, we can apply the second order topology design Algorithm 2 to solve a class of inverse reconstruction problem. See also the related works [14, 60, 71, 72, 78, 103].

10.4.1 Inverse Gravimetry Problem

The inverse gravimetry problem consists in reconstructing the mass distribution in a geometrical domain from measurements of the gravity force on the boundary [136]. Based on [68], the inverse problem is reformulated as a topology optimization problem, where the support of the mass distribution is the unknown variable. The Kohn–Vogelius functional, which measures the misfit between the solutions of two auxiliary problems, one containing information about the boundary measurement and the other for the boundary excitation, is minimized. The Newtonian potential is used to complement the unavailable information about the hidden boundary. The resulting topology optimization algorithm is based on an analytic formula for the variation of the Kohn–Vogelius functional with respect to a class of mass distributions consisting of a finite number of ball-shaped trial anomalies.

In particular, let us consider an open bounded domain $\Omega \subset \mathbb{R}^3$, with Lipschitz boundary $\partial\Omega$. We introduce the boundary $\Gamma_M \subset \partial\Omega$ where the measurements are taken, and $\Gamma = \partial\Omega \setminus \Gamma_M$ is the remaining (hidden) boundary, where there is no information. We assume that the unknown density b^* of the medium belongs to the following set of admissible sources:

$$C_\gamma(\Omega) := \left\{ \varphi \in L^\infty(\Omega) : \varphi = \sum_{i=1}^{N} \gamma_i \mathbb{1}_{\omega_i} \right\}, \tag{10.9}$$

where the sets $\omega_i \subset \Omega$, with $i = 1, \ldots, N$, are such that $\omega_i \cap \omega_j = \varnothing$ for $i \neq j$. In addition, $\mathbb{1}_{\omega_i}$ is the characteristic function of ω_i. Finally, $\gamma_i \in \mathbb{R}_+$ are the contrasts with respect to the density of the background, which are assumed to be known. From $b^* \in C_\gamma(\Omega)$ we define the associated potential as [138]

$$u(x) = \int_\Omega K(x, y) b^*(y)\, dy\ , \tag{10.10}$$

where the kernel $K(x, y)$ is given by

$$K(x, y) = \frac{1}{4\pi \|x - y\|}. \tag{10.11}$$

The boundary measurements (U, Q) are defined as

$$U := u_{|\Gamma_M} \quad \text{and} \quad Q := -\partial_n u_{|\Gamma_M}\ , \tag{10.12}$$

where n is the outward unit normal vector to Ω.

Now, the inverse gravimetry problem we are dealing with reads: given $Q \in H^{-1/2}(\Gamma_M)$ and $U \in H^{1/2}(\Gamma_M)$, find the unknown source $b^* \in C_\gamma(\Omega)$ such that there exists $u^* \in H^1(\Omega)$ satisfying the following overdetermined boundary value problem:

$$\begin{cases} -\Delta u^* = b^* & \text{in } \Omega\ , \\ \left.\begin{aligned} u^* &= U \\ -\partial_n u^* &= Q \end{aligned}\right\} & \text{on } \Gamma_M\ , \end{cases} \tag{10.13}$$

The inverse problem (10.13) is clearly ill-posed [135]. However, since the contrasts γ_i, $i = 1, \ldots N$, in (10.9) are assumed to be known, problem (10.13) can be written as a topology optimization problem with respect to the sets $\omega = \bigcup_{i=1}^N \omega_i$. Therefore, we introduce the following topology optimization problem based on minimization of the Kohn–Vogelius criterion [153] with respect to the support ω:

$$\underset{\omega \subset \Omega}{\text{Minimize}}\ \mathcal{J}_\omega(u^D, u^N) = \int_\Omega (u^D(\omega) - u^N(\omega))^2, \tag{10.14}$$

which will be solved by using first and second order topological derivatives. The auxiliary functions $u^D = u^D(\omega)$ and $u^N = u^N(\omega)$ are solutions to the following boundary value problems:

$$\begin{cases} -\Delta u^D = b_\omega & \text{in } \Omega\ , \\ u^D = U & \text{on } \Gamma_M\ , \\ u^D = u^T & \text{on } \Gamma\ , \end{cases} \quad \text{and} \quad \begin{cases} -\Delta u^N = b_\omega & \text{in } \Omega\ , \\ -\partial_n u^N = Q & \text{on } \Gamma_M\ , \\ u^N = u^T & \text{on } \Gamma\ . \end{cases} \tag{10.15}$$

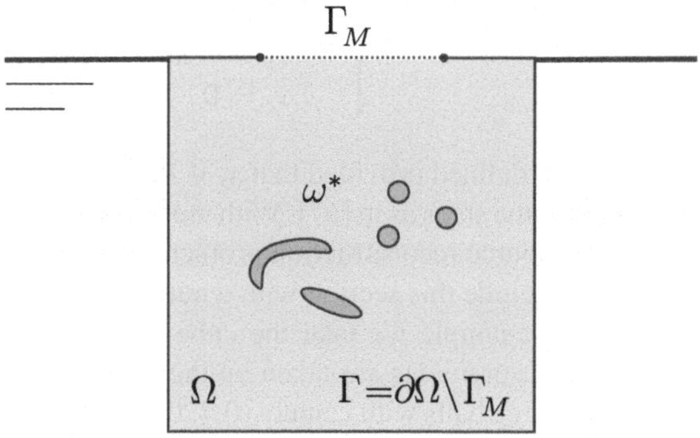

Fig. 10.3 The inverse gravimetry problem

where $b_\omega \in C_\gamma(\Omega)$ and the Newtonian potential

$$u^T(x) = \int_\Omega K(x, y) b_\omega(y)\, dy \tag{10.16}$$

is used to complement the information on the hidden boundary Γ, with the kernel $K(x, y)$ given by (10.11). Note that the domain Ω and the part Γ of $\partial\Omega$ do not represent physical quantities and are introduced to get a meaningful mathematical model. Actually, the inverse gravimetry problem may e.g. be defined in the whole half-space $\mathbb{R}^2 \times (-\infty, 0)$, as represented in Fig. 10.3. The only constraint on Ω is that it has to be large enough to contain any possible anomaly, since (10.16) is correct only if this requirement is satisfied. In the following we assume that any possible anomaly is in Ω.

We consider $\omega = \varnothing$ as initial guess, so that $b_{\omega|_{\omega=\varnothing}} = 0$. The domain Ω is perturbed by the nucleation of N ball-shaped anomalies of radii ε_i and centers at $x_i \in \Omega$, endowed with contrasts γ_i, $i = 1, \ldots, N$. From these elements, $\alpha = (|B_{\varepsilon_1}(x_1)|, \ldots, |B_{\varepsilon_N}(x_N)|)$ and $\mathscr{E}(\varepsilon) \equiv 0$ in (10.3), so there is no remainder in this case. In addition, the entries of the vector $d(\xi) \in \mathbb{R}^N$ and of the matrix $H(\xi) \in \mathbb{R}^N \times \mathbb{R}^N$ are defined as

$$d_i = 2\gamma_i \int_\Omega (u^D - u^N) h_i \quad \text{and} \quad H_{ij} = \gamma_i \gamma_j \int_\Omega h_i h_j \,. \tag{10.17}$$

Finally, the auxiliary functions $h_i(x)$ are solutions to the following boundary value problems depending on x_i:

$$\begin{cases} -\Delta h_i = 0 \ \text{ in } \Omega \,, \\ -\partial_n h_i = g_i \ \text{ on } \Gamma_M \,, \\ h_i = 0 \ \text{ on } \Gamma \,, \end{cases} \tag{10.18}$$

where $g_i = \partial_n w_i$ on Γ_M with $w_i(x) = K(x, x_i) + v_i(x)$ and v_i solving

$$\begin{cases} -\Delta v_i = 0 & \text{in } \Omega , \\ v_i = -K(x, x_i) & \text{on } \Gamma_M , \\ v_i = 0 & \text{on } \Gamma , \end{cases} \tag{10.19}$$

which is well defined provided that $x_i \notin \partial\Omega$. The derivation of the above equations follows the same steps as in [67]. With the above elements, we can apply Algorithm 2 to solve the source reconstruction problem (10.14).

Let us conclude this section with a numerical experiment in three spatial dimensions. In the example we take the cube $\Omega = (0, 1) \times (0, 1) \times (0, 1)$. The partial boundary measurements are taken on the side$\Gamma_M = \{x = 1\} \cap \overline{\Omega}$. We are going to reconstruct three balls with centers $(0.7, 0.7, 0.4)$, $(0.6, 0.3, 0.3)$, $(0.5, 0.4, 0.8)$ and radii 0.17, 0.2, 0.15 respectively. The intensities γ_i are 1 for $i = 1, \ldots, 3$. We use synthetic measurements without noise in the data. Since in the set of ball-shaped anomalies there is no remainder in the expansion (10.3), we observe that Algorithm 2 reconstructs the positions and sizes of the three balls exactly. The result is shown in Fig. 10.4.

10.4.2 Pointwise Source Reconstruction Problem

We consider a benchmark example concerning an inverse potential problem in two spatial dimensions. Let $\Omega \subset \mathbb{R}^2$ be an open bounded domain with Lipschitz boundary $\partial\Omega$. The inverse problem we are dealing with consists in determining the unknown pointwise source $b^* \in C_\delta(\Omega)$ from the Cauchy data U and Q in the following elliptic boundary value problem:

$$\begin{cases} -\Delta u^* = b^* & \text{in } \Omega, \\ \left.\begin{aligned} u^* &= U \\ -\partial_n u^* &= Q \end{aligned}\right\} & \text{on } \partial\Omega. \end{cases} \tag{10.20}$$

where the set $C_\delta(\Omega)$ is defined as

$$C_\delta(\Omega) = \left\{ \varphi : \Omega \mapsto \mathbb{R} \mid \varphi(x) = \sum_{i=1}^{N} \alpha_i \delta(x - x_i) \right\}, \tag{10.21}$$

with $\alpha_i \in \mathbb{R} \setminus \{-\infty, +\infty\}$ and $x_i \in \Omega$, for $i = 1, \ldots, N$. Therefore, the unknown source $b^* \in C_\delta(\Omega)$ can be represented as follows:

$$b^*(x) = \sum_{i=1}^{N^*} \alpha_i^* \delta(x - x_i^*). \tag{10.22}$$

Thus, solving the above inverse potential problem in $C_\delta(\Omega)$ means to find N^*, α_i^* and x_i^*, which denote the number, intensities and locations of the unknown pointwise

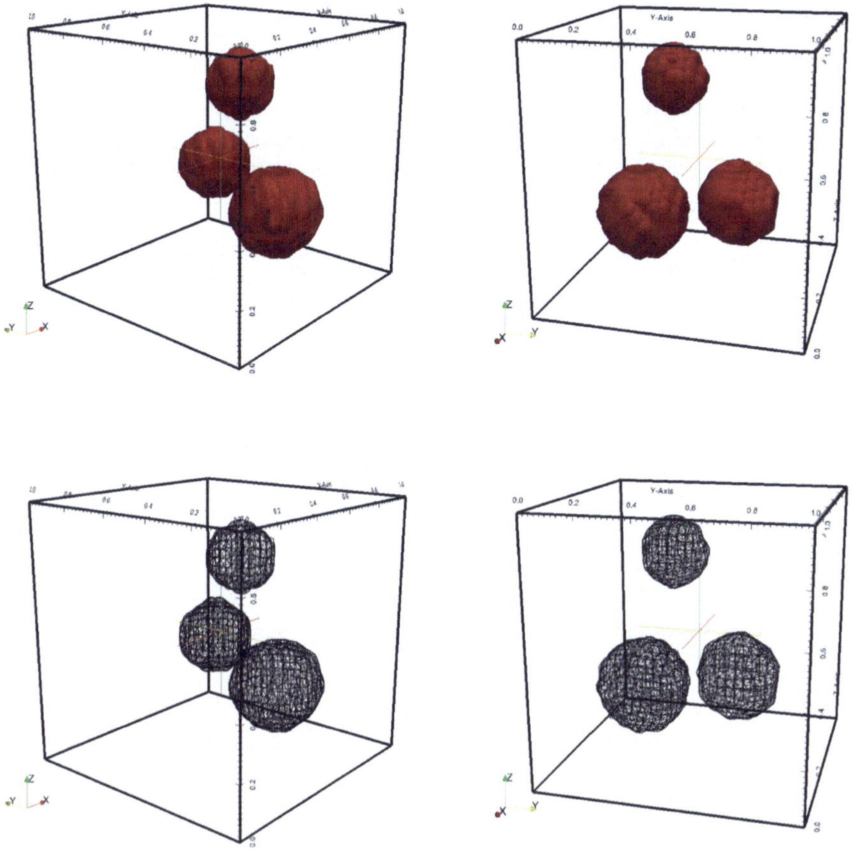

Fig. 10.4 True anomalies (first row) and corresponding reconstructions (second row) for the gravimetry inverse problem [68]

sources, respectively. Let us introduce the following functional based on the Kohn–Vogelius criterion [153]:

$$\underset{b\in C_\delta(\Omega)}{\text{Minimize}}\ \mathscr{J}(u^D,u^N)=\int_\Omega (u^D(\omega)-u^N(\omega))^2, \tag{10.23}$$

where the functions u^D and u^N are solutions to the following auxiliary problems:

$$\begin{cases} -\Delta u^D = b & \text{in } \Omega, \\ \quad u^D = U & \text{on } \partial\Omega. \end{cases} \quad \text{and} \quad \begin{cases} -\Delta u^N = b + c & \text{in } \Omega, \\ -\partial_n u^N = Q & \text{on } \partial\Omega, \\ \int_\Omega u^N = \int_\Omega u^D, & \end{cases} \tag{10.24}$$

where $b \in C_\delta(\Omega)$ is a given source, representing an initial guess. In addition, the compatibility constant c is given by

$$c = \frac{1}{|\Omega|}\left(\int_{\partial\Omega} Q - \int_{\Omega} b\right), \tag{10.25}$$

where $|\Omega|$ is the Lebesgue measure of Ω.

Let us now perturb the source b by introducing a number N of pointwise sources with arbitrary locations $x_i \in \Omega$ and intensities α_i, with $i = 1, \ldots, N$. The perturbed source $b_\delta \in C_\delta(\Omega)$ is defined as follows:

$$b_\delta(x) = b(x) + \sum_{i=1}^{N} \alpha_i \delta(x - x_i). \tag{10.26}$$

For this class of perturbations, the vector $d(\xi) \in \mathbb{R}^N$ and the matrix $H(\xi) \in \mathbb{R}^N \times \mathbb{R}^N$ in (10.6) can be obtained explicitly:

$$d_i = \int_{\Omega} h_i(u^D - u^N) \quad \text{and} \quad H_{ij} = \int_{\Omega} h_i h_j, \tag{10.27}$$

where the auxiliary functions h_i are solutions to

$$\begin{cases} -\Delta h_i = |\Omega|^{-1} & \text{in } \Omega, \\ -\partial_n h_i = \partial_n v_i & \text{on } \partial\Omega, \\ \int_{\Omega} h_i = 0, \end{cases} \quad \text{with} \quad \begin{cases} -\Delta v_i = \delta_i & \text{in } \Omega, \\ v_i = 0 & \text{on } \partial\Omega, \end{cases} \tag{10.28}$$

with $\delta_i(x) := \delta(x - x_i)$. The derivations of the above formulas can be found in [181], for instance.

Now, we have all elements to apply Algorithm 2 to solve the pointwise source reconstruction problem. In fact, let $\Omega = (-0.5, 0.5) \times (-0.5, 0.5)$. The target consists of three pointwise sources shown in Fig. 10.5a, where the radius of each ball represents the associated intensities. Here, we impose $U = 0$ and observe Q on $\partial\Omega$. The initial guess $b \in C_\delta(\Omega)$ is identically zero, that is, $b = 0$ in Ω. The result is shown in Fig. 10.5b, where the locations and intensities are perfectly reconstructed.

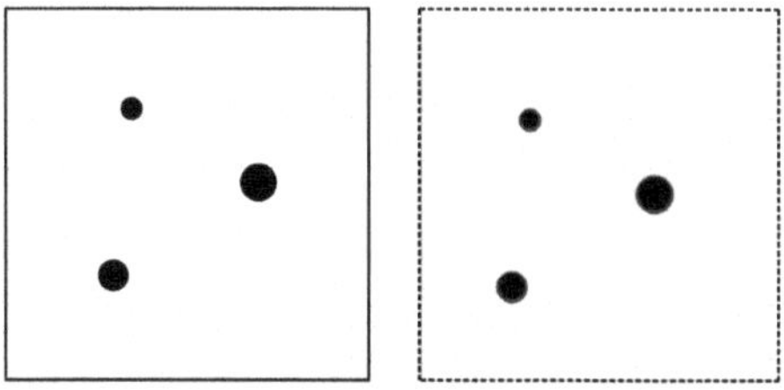

(a) target (b) result

Fig. 10.5 Target and result for the pointwise source reconstruction problem [181]

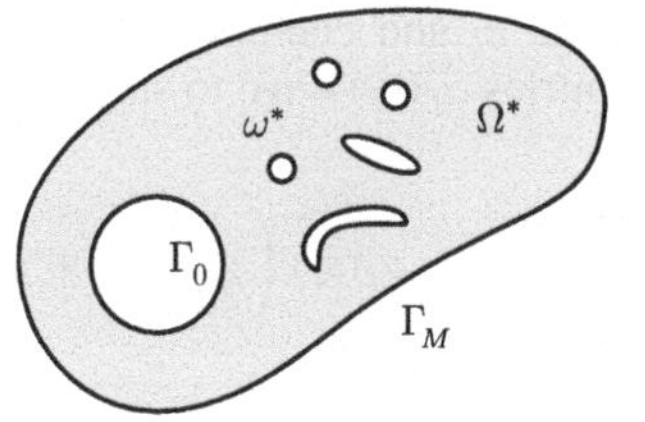

Fig. 10.6 The inverse obstacle problem

10.4.3 Obstacle Reconstruction Problem

Let $\mathscr{D} \subset \mathbb{R}^2$ be an open bounded domain with smooth boundary Γ_M. We introduce a subset Ω of $\mathscr{D}$ such that $\Omega = \mathscr{D} \setminus \overline{\omega_0}$, with $\omega_0 \Subset \mathscr{D}$. The boundary of Ω is split into two disjoint parts Γ_M and Γ_0, where Γ_0 is the boundary of the hole ω_0. Let us consider the domain $\Omega^* = \Omega \setminus \overline{\omega^*}$, where $\omega^* \Subset \Omega$ represents a number $N^* \in \mathbb{N}$ of unknown holes (obstacles) in Ω^*. The boundary of Ω^* is split into three disjoint subsets Γ_M, Γ_0 and $\partial\omega^*$, where $\partial\omega^*$ denotes the boundaries of the N^* obstacles ω^*. See sketch in Fig. 10.6. The inverse problem we are dealing with is to find ω^* such that the following overdetermined boundary value problem is satisfied:

$$\left\{ \begin{array}{rll} \Delta u^* &= 0 & \text{in } \Omega^*, \\ u^* &= 0 & \text{on } \Gamma_0, \\ u^* &= 0 & \text{on } \partial\omega^*, \\ \left.\begin{array}{r} u^* = U \\ -\partial_n u^* = Q \end{array}\right\} & & \text{on } \Gamma_M, \end{array} \right. \tag{10.29}$$

where U and Q are the Cauchy data on Γ_M. We assume that the flux Q is imposed while the potential U is measured.

Since the inverse problem (10.29) is formulated as an ill-posed and overdetermined boundary value problem, the idea is to rewrite it as a topology optimization problem:

$$\underset{\Omega \subset \mathscr{D}}{\text{Minimize}} \quad \mathscr{J}_\Omega(u) = \int_{\Gamma_M} (u - U)^2. \tag{10.30}$$

Some terms above require explanation. The shape functional $\mathscr{J}_\Omega(u)$ measures the misfit between the boundary measurement U and the trace on Γ_M of the solution $u = u(\Omega)$ to the following auxiliary boundary value problem depending on the boundary data Q:

$$\left\{ \begin{array}{rl} \Delta u = 0 & \text{in } \Omega, \\ u = 0 & \text{on } \Gamma_0, \\ -\partial_n u = Q & \text{on } \Gamma_M. \end{array} \right. \tag{10.31}$$

The topological derivative concept is used to solve problem (10.30), which has been specifically designed to deal with such a topology optimization problem. In particular, the domain Ω is perturbed by the nucleation of N ball-shaped holes of radii ε_i and centers at $x_i \in \Omega$, with $i = 1, \ldots, N$. In this case, $\alpha = \alpha(\varepsilon)$, whose entries $\alpha_i(\varepsilon_i)$ have to satisfy the constraint

$$u(x_i) - \alpha_i(\varepsilon_i)\left(\frac{1}{2\pi}\ln\varepsilon_i + g_i(x_i)\right) - \sum_{\substack{j=1\\ j\neq i}}^{N} \alpha_j(\varepsilon_j)\left(\frac{1}{2\pi}\ln\|x_i - x_j\| + g_j(x_i)\right) = 0, \tag{10.32}$$

with each g_j solving the following auxiliary boundary value problem:

$$\begin{cases} \Delta g_j = 0 & \text{in } \Omega, \\ \partial_n g_j = \partial_n \phi_j & \text{on } \Gamma_M, \\ g_j = \phi_j & \text{on } \Gamma_0, \end{cases} \tag{10.33}$$

where ϕ_j is the fundamental solution for the Laplacian in two spatial dimensions,

$$\phi_j(x) = \frac{1}{2\pi}\ln\|x - x_j\|, \quad \forall x \in \Omega. \tag{10.34}$$

Finally, the entries of the vector $d(\xi) \in \mathbb{R}^N$ and of the matrix $H(\xi) \in \mathbb{R}^N \times \mathbb{R}^N$ in (10.6) are defined by

$$d_i = \int_{\Gamma_M} (u - U)G_i \quad \text{and} \quad H_{ij} = \int_{\Gamma_M} G_i G_j, \tag{10.35}$$

where the auxiliary functions G_j are solutions to

$$\begin{cases} -\Delta G_j = \delta_j & \text{in } \Omega, \\ \partial_n G_j = 0 & \text{on } \Gamma_m, \\ G_j = 0 & \text{on } \Gamma_0, \end{cases} \tag{10.36}$$

with $\delta_j := \delta(x - x_j)$ denoting the Dirac delta distributions. The derivation of the above formulas can be found in [225], for instance.

Remark 10.1 Note that $g_j(x) = g_j(x; x_j)$, because $\phi_j(x) = \phi_j(x; x_j)$ depends on the point x_j where the hole is nucleated. This means that the function $g_j(x)$ depends on x_j. Therefore, the complexity of the computation of all g_j may be very high. However, we will show through a numerical experiment that this term is crucial for solving the topology optimization problem (10.30) or equivalently the inverse problem (10.29); which can be seen as the main theoretical contribution from [225].

To fix ideas, let us present a simple example. We consider the domain Ω given by a disk centered at (0, 0) and with unit radius. In addition, Ω has a hole ω_0 centered at (−0.5, 0) and with radius 0.3. The target domain Ω^* has three hidden circular obstacles represented by black circles in Fig. 10.7a. A uniform flux $Q = 1$ is applied on Γ_M, where the associated potential U is measured. The resulting reconstruction obtained from Algorithm 2 is represented by black circles in Fig. 10.7b.

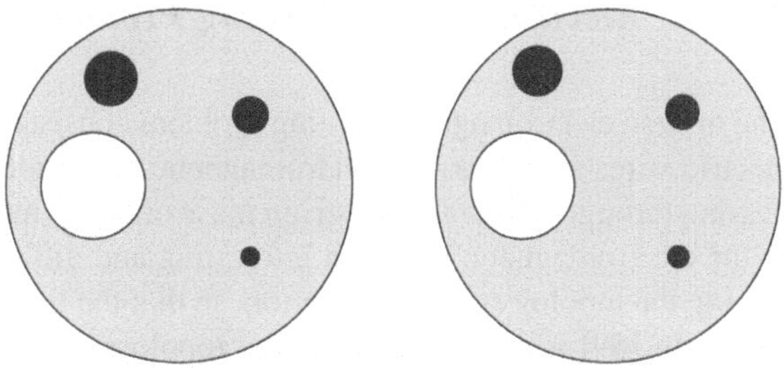

(a) target (b) result

Fig. 10.7 Target and result for the inverse obstacle problem [225]

10.4.4 Inverse Conductivity Problem

The electrical impedance tomography (EIT) problem consists in determining the distribution of the electrical conductivity of a medium subject to a set of current fluxes, from measurements of the corresponding electrical potentials on its boundary. EIT is probably the most studied inverse problem since the fundamental works by Calderón from the eighties. It has many relevant applications in medicine (detection of tumors), geophysics (localization of mineral deposits) and engineering (detection of corrosion in structures). Important contributions can be found in [51, 71, 78, 103, 152, 154, 155]. The stability and resolution analysis for a (first-order) topological derivative based imaging functional in the context of Helmholtz equation is known [13]. However, such an analysis is missing for the inverse conductivity problem. Therefore, the second-order topological derivative concept starts to play an important role in the context of inverse reconstruction problems. In particular, it has been successfully applied for solving a class of EIT problems [55, 101, 130]. In the paper [55] a higher order expansion of a tracking-type shape functional with respect to a number of arbitrary shaped inclusion is derived, all of them controlled by the same small parameter. The resulting expansion is used to reconstruct a single circular or elliptical inclusion from partial boundary measurement. In [130] a second-order topological expansions of a tracking-type shape functional is also considered with respect to several circular inclusions of uniform sizes. The resulting expansion is used to initialize a standard level set method. In the approach proposed in [101], the anomalies are approximated by a finite number of ball-shaped trial inclusions of different sizes. In addition, the adjoint method is evoked a posteriori, after obtaining the associated sensitivities, allowing to derive a simpler representation for the resulting expansion. These two ingredients were crucial in the development of a novel non-iterative reconstruction algorithm, which represents the main contribution of [101] with respect to [55, 130]. These ideas are fully developed in Chap. 11.

10.4.5 *Electromagnetic Casting Problem*

The inverse electromagnetic casting problem consists in looking for a suitable set of electric wires such that the electromagnetic field induced by an alternating current passing through them makes a given mass of liquid metal assume a shape according to the electromagnetic field. An interesting and difficult problem consists in determining the topology of such inductors so that the liquid metal acquires a predefined shape. In [69] a new method for the topology design of inductors in the inverse electromagnetic casting problem has been proposed. See also [70]. It relies on the topological derivative concept. The basic idea is to reformulate the inverse electromagnetic casting problem as a topology optimization problem, where the associated shape functional is minimized with respect to a set of ball-shaped inductors. Based on the theoretical result, a topology design algorithm of inductors has been devised. In those papers, several numerical examples are presented showing that the technique is effective.

10.5 Future Developments

In this chapter a second order method has been presented, together with a set of applications in the context of inverse reconstruction problems. The general idea consists in rewrite the inverse problem as a topology optimization problem, where a shape functional measuring the misfit between the boundary measurements and the solution obtained from the model is expanded with respect to a set of ball-shaped anomalies. The resulting expansion is then truncated up to the second order term, leading to a quadratic and strictly convex form with respect to the parameters under consideration. Therefore, the truncated expansion has been used to devise a novel non-iterative reconstruction algorithm based on a simple optimization step. As a result, the reconstruction process becomes very robust with respect to noisy data and also independent of any initial guess. Finally, these ideas were used for solving a wide class of inverse reconstruction problems. Since the proposed method can approximate accurately the unknown set of hidden anomalies by several balls, it can be used for supplying a good initial guess for more complex iterative approaches, such as the ones based on level-sets methods [163], for instance. We would like to stress however that the methodology here presented is quite recent. Therefore, it is still not clear yet the best way to use the second order topological derivative concept. For future development of this new branch of shape-topological second order methods we highlight the following:

1. There is no stability and resolution analysis for the second order topological derivative based imaging functional. To demonstrate these properties in general, or at least for some particular inverse reconstruction problems, would be an important contribution.

2. Replacing the Newton method based on second order topological derivatives by a family of quasi-Newton methods seems to deserve investigation.
3. The bottleneck of the reconstruction Algorithm 2 relies on its complexity given by formula (10.8) for a high number of trial balls N, with $M \gg N$. Some insight on how to deal with the complexity issue can be found in [181]. However, it can be seen as an interesting problem, including the question of how to efficiently use higher order expansions.
4. Extension of second order topological derivatives to the context of defect detection in elasticity or elastodynamics can also be investigated. However, technical difficulties are to be expected.
5. Applications of the second order method to other classes of problems, including all those presented in Chap. 6, can also be seen as an interesting and sometimes difficult research topic. In fact, it requires the development of new asymptotic formulas. In addition, many of the problems have non-unique solution, so that the Hessian matrix could degenerate. Therefore, new regularization strategies would be required to get well-posedness.
6. We mention that the open problem proposed in [137, pp. 126, Problem 4.2] still remains unsolved. We believe that higher order topological asymptotic expansions may give some insight on how to solve this open problem by exploring the properties of the associated Hessian matrix. See the recent publications [99, 100], for instance.
7. The uniqueness result for the gravimetry problem presented in Sect. 10.4.1 can be found in [135]. However, it ensures uniqueness for one single star-shaped geological anomaly. In contrast, according to the numerical experiment presented in Fig. 10.4, it is clear that we can reconstruct more than one single anomaly. Therefore, extension of the uniqueness result by considering a set of multiply connected anomalies deserves investigation. It seems that this result should be true provided that the complement $\Omega \setminus \omega^*$ is simply connected.

Chapter 11
The Electrical Impedance Tomography Problem

In this chapter the Electrical Impedance Tomography (EIT) is adopted as model problem, which consists in finding the number, size, shape and location of a set of hidden anomalies inside a body from total or partial measurements of the electrical potential on the boundary of the body. Since the EIT problem is written in the form of an overdetermined boundary value problem, the idea is to rewrite it as a topology optimization problem. In particular, a shape functional measuring the misfit between the boundary measurements and the electrical potentials obtained from the model is minimized with respect to a set of ball-shaped anomalies by using the concept of topological derivatives. It means that the objective functional is expanded and then truncated up to the second order term, leading to a quadratic and strictly convex form with respect to the parameters under consideration. Thus, a trivial optimization procedure allows for devising a non-iterative second order reconstruction algorithm. As a result, the reconstruction process becomes very robust with respect to noisy data and independent of any initial guess. Finally, in order to show the effectiveness of the devised reconstruction algorithm, some numerical experiments into two spatial dimensions are presented, taking into account partial boundary measurements and noisy data. This chapter is based on the paper by Andrey Dione Ferreira and Antonio André Novotny [101].

11.1 The Inverse Conductivity Problem

Let us consider a domain $\Omega \subset \mathbb{R}^2$ with Lipschitz continuous boundary $\partial\Omega$. The domain Ω represents a body endowed with the capability of conducting electricity. Its electrical conductivity coefficient is denoted by $k^*(x) \geq k_0 > 0$, with $x \in \Omega$ and $k_0 \in \mathbb{R}_+$. If the body Ω is subjected to a given electric flux Q on $\partial\Omega$, then the resulting electric potential u in Ω is observed on a part of the boundary $\Gamma_m \subset \partial\Omega$. The

A. A. Novotny et al., *Applications of the Topological Derivative Method*, Studies in Systems, Decision and Control 188,
https://doi.org/10.1007/978-3-030-05432-8_11

objective is to reconstruct the electrical conductivity k^* over Ω from a given boundary measurement $u^*_{|\Gamma_m} = U$, solution of the following overdetermined boundary value problem

$$\begin{cases} \operatorname{div}[q(u^*)] = 0 & \text{in } \Omega, \\ q(u^*) = -k^* \nabla u^*, & \\ q(u^*) \cdot n = Q & \text{on } \partial\Omega, \\ u^* = U & \text{on } \Gamma_m. \end{cases} \tag{11.1}$$

Without loss of generality, we are considering only one boundary measurement U on Γ_m. The extension to several boundary measurements is trivial.

An important feature of the human body is that the electrical conductivity can be approximated by a piecewise constant function representing different tissues. The electrical conductivity of the muscles, lungs, bones and blood are respectively given by 8.0, 1.0, 0.06 and 6.7 millisiemens (mS), for instance. Thus, we assume that the unknown electrical conductivity k^* we are looking for belongs to the following set

$$\mathcal{C}_\gamma(\Omega) := \left\{ \varphi \in L^\infty(\Omega) : \varphi = k \left(\mathbb{1}_\Omega - \sum_{i=1}^N (1-\gamma_i) \mathbb{1}_{\omega_i} \right) \right\}, \tag{11.2}$$

where $k \in \mathbb{R}_+$ is the electrical conductivity of the background. When the conductivity k depends on the frequency, see [2]. The sets $\omega_i \subset \Omega$, with $i = 1, \ldots, N$, are such that $\omega_i \cap \omega_j = \emptyset$, for $i \neq j$. In addition, $\mathbb{1}_\Omega$ and $\mathbb{1}_{\omega_i}$ are used to denote the characteristics functions of Ω and ω_i, respectively. Finally, $\gamma_i \in \mathbb{R}_+$ are the contrasts with respect to the electrical conductivity of the background k. From these elements, the inverse problem we are dealing with can be stated as:

Problem 11.1 Let $Q \in H^{-1/2}(\partial\Omega)$ be a given Neumann excitation, then find $k^* \in \mathcal{C}_\gamma(\Omega)$ from observations of the field U on $\Gamma_m \subset \partial\Omega$, such that $u^*[k^*] \in H^1(\Omega)$ satisfies (11.1). See sketch in Fig. 11.1.

We assume that each ω_i is measurable and simply connected. We also assume that the values of the electrical conductivity of the background k and the associated contrasts γ_i are known (see counter-example at the end of this section). From these assumptions, Problem 11.1 can be written as a topology optimization problem with respect to the sets ω_i, for $i = 1, \ldots, N$. In fact, let us introduce the unknown set $\omega^* = \bigcup_{i=1}^{N^*} \omega_i^* \subset \Omega$, where N^* is the number of anomalies we are looking for. Therefore

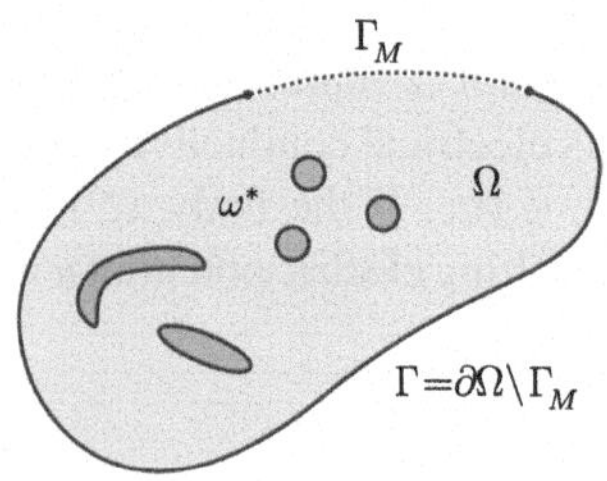

Fig. 11.1 The electrical impedance tomography problem

$$k^*(x) = \begin{cases} k & \text{if } x \in \Omega \setminus \omega^*, \\ \gamma_i k & \text{if } x \in \omega_i^*,\ i = 1, \ldots, N^*. \end{cases} \tag{11.3}$$

Now, let us introduce the following auxiliary Neumann boundary value problem: Find u, such that

$$\begin{cases} \text{div}[q(u)] = 0 & \text{in } \Omega, \\ q(u) = -k_\omega \nabla u, & \\ q(u) \cdot n = Q & \text{on } \partial\Omega, \\ \int_{\partial\Omega} Q = 0, & \\ \int_{\Gamma_m} u = \int_{\Gamma_m} U, & \end{cases} \tag{11.4}$$

where Q and U are the boundary excitation and boundary measurement, respectively and $k_\omega \in C_\gamma(\Omega)$ is constant by parts, characterized by a set $\omega \subset \Omega$.

Finally, we introduce the following shape functional measuring the misfit between the boundary measurement U and the solution $u = u(\omega)$ of (11.4) evaluated on Γ_m, namely

$$\mathscr{J}_\omega(u) = \int_{\Gamma_m} (u - U)^2. \tag{11.5}$$

Since the electrical conductivity of the background k and the associated contrasts γ_i are known by assumption, then solving Problem 11.1 is equivalent to solve the following topology optimization problem with respect to the support ω of the anomalies

$$\underset{\omega \subset \Omega}{\text{Minimize}}\ \mathscr{J}_\omega(u), \quad \text{subject to (11.4)}. \tag{11.6}$$

The problem of finding $k^* \in C_\gamma(\Omega)$ for a given Dirichlet excitation U on $\partial\Omega$, from observations of the flux Q on $\Gamma_m \subset \partial\Omega$, by using similar optimization approach can be found in [20]. The minimization problem (11.6) we are dealing with is based on the following paradigm: we know what we are looking for, but we do not know where they are. In fact, we cannot reconstruct both the topology ω^* and the contrasts $\gamma_i, i = 1, \ldots, N^*$.

Remark 11.1 In the case of the EIT problem in general a number M of boundary measurements is available, which are easily obtained by combining different pairs of injection and draining electrodes. See sketch in Fig. 11.2, for instance. However, for the sake of simplicity and without lost of generality, in this chapter all derivations are presented by taking into account just one single boundary measurement. Their extensions for a number $M > 1$ of measurements is trivially obtained after sum-up the sensitivities associated with each individual boundary measurement.

Fig. 11.2 Model problem

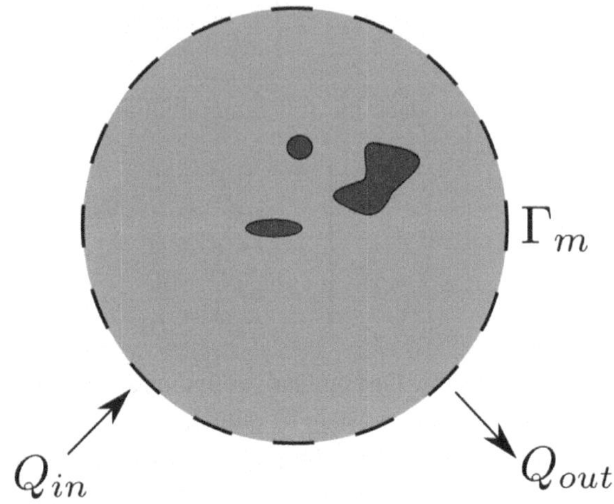

11.2 Topological Asymptotic Expansion

In this chapter we deal with the reconstruction of $k^* \in C_\gamma(\Omega)$ from total or partial boundary measurement using the topological sensitivity analysis concept [219]. Therefore, let us consider $\omega = \emptyset$ and $u_0 = u_{|_{\omega=\emptyset}}$ solution of

$$\begin{cases} \operatorname{div}[q(u_0)] = 0 & \text{in } \Omega, \\ q(u_0) = -k\nabla u_0, & \\ q(u_0) \cdot n = Q & \text{on } \partial\Omega, \\ \int_{\partial\Omega} Q = 0, & \\ \int_{\Gamma_m} u_0 = \int_{\Gamma_m} U, & \end{cases} \tag{11.7}$$

where k is a constant that represents the electrical conductivity of the background. In this particular case, the following notation for the shape function is introduced

$$\mathcal{J}_0(u_0) = \int_{\Gamma_m} (u_0 - U)^2. \tag{11.8}$$

Let us perturb the domain Ω by nucleating – simultaneously – a number N of circular inclusions $B_{\varepsilon_i}(x_i)$ with contrast γ_i, $i = 1, \ldots, N$. We assume that $B_{\varepsilon_i}(x_i) \subset \Omega$ is a ball with center at $x_i \in \Omega$ and radius ε_i, such that $B_{\varepsilon_i}(x_i) \cap B_{\varepsilon_j}(x_j) = \emptyset$ for $i \neq j$. We introduce the notations $\xi = (x_1, \ldots, x_N)$ and $\varepsilon = (\varepsilon_1, \ldots, \varepsilon_N)$, whether necessary. The topologically perturbed counterpart of the shape functional is defined as follows

$$\mathcal{J}_\varepsilon(u_\varepsilon) = \int_{\Gamma_m} (u_\varepsilon - U)^2, \tag{11.9}$$

where u_ε is solution of the following boundary value problem

$$\begin{cases} \operatorname{div}[q_\varepsilon(u_\varepsilon)] = 0 & \text{in } \Omega, \\ q_\varepsilon(u_\varepsilon) = -\gamma_\varepsilon k \nabla u_\varepsilon, & \\ q_\varepsilon(u_\varepsilon) \cdot n = Q & \text{on } \partial\Omega, \\ \int_{\partial\Omega} Q = 0, & \\ \int_{\Gamma_m} u_\varepsilon = \int_{\Gamma_m} U, & \\ [\![u_\varepsilon]\!] = 0 & \text{on } \bigcup_{i=1}^N \partial B_{\varepsilon_i}(x_i), \\ [\![q_\varepsilon(u_\varepsilon)]\!] \cdot n = 0 & \text{on } \bigcup_{i=1}^N \partial B_{\varepsilon_i}(x_i), \end{cases} \tag{11.10}$$

with the contrast defined as

$$\gamma_\varepsilon = \gamma_\varepsilon(x) = \begin{cases} 1, & \text{if } x \in \Omega \setminus \bigcup_{i=1}^N B_{\varepsilon_i}(x_i) \\ \gamma_i, & \text{if } x \in B_{\varepsilon_i}(x_i). \end{cases} \tag{11.11}$$

11.2.1 Asymptotic Expansion of the Solution

General results for the asymptotic expansions of solutions in singularly perturbed domains were originally considered in [134, 184]. When perturbations are included in Ω, discrepancies over the inclusions $B_{\varepsilon_i}(x_i)$ in problem (11.7) appear. The idea is to introduce boundary layers which compensate for such discrepancies. Thus, let us consider the following *ansätz* for the asymptotic expansion of u_ε

$$u_\varepsilon(x) = u_0(x) + \sum_{i=1}^N \left(w_i(x/\varepsilon_i) + \varepsilon_i^2 \tilde{u}_i(x) + \tilde{w}_i(x/\varepsilon_i) + \varepsilon_i^4 \tilde{\tilde{u}}_i(x) \right)$$
$$+ \sum_{i=1}^N \sum_{j \neq i}^N \left(w_i^j(x/\varepsilon_i) + \varepsilon_i^2 \varepsilon_j^2 u_i^j(x) \right) + \tilde{\tilde{u}}_\varepsilon(x). \tag{11.12}$$

Before continue, let us give a rough explanation of each term in the above expansion. The boundary layers w_i are introduced to compensate for the first and second order terms of the Taylor's expansion of ∇u_0 around x_i. The problem associate with $\tilde{u}_i$ compensates for the discrepancy introduced on $\partial\Omega$ by one term of w_i. The boundary layers $\tilde{w}_i$ compensate for the first and third term of the Taylor's expansion of $\nabla \tilde{u}_i$ and ∇u, respectively. There still discrepancies left on $\partial\Omega$ by one term of each set of boundary layers w_i and $\tilde{w}_i$, which are compensate by $\tilde{\tilde{u}}_i$. The terms w_i^j and u_i^j are introduced to take into account interactions between different inclusions. Finally, $\tilde{\tilde{u}}_\varepsilon$ compensate for all remainder discrepancies. Each term of the *ansätz* (11.12) are now explicitly defined. In what follows, the notation $\nabla^n \varphi(y)(x - x_i)^n$ represents the derivative of order n of a function φ in the direction $(x - x_i)$ evaluated at y.

We start with the boundary layers $w_i(x/\varepsilon_i)$, for $i = 1, \ldots, N$, which are solutions of

$$\begin{cases} \operatorname{div}[q_\varepsilon(w_i)] = 0 & \text{in } \Xi_{\varepsilon_i}, \\ q_\varepsilon(w_i) = -\gamma_{\varepsilon_i} k \nabla w_i & \text{in } \mathbb{R}^2, \\ w_i \to 0 & \text{at } \infty, \\ [\![w_i]\!] = 0 & \text{on } \partial B_{\varepsilon_i}(x_i), \\ [\![q_\varepsilon(w_i)]\!] \cdot n = k(1-\gamma_i)\left(\nabla u_0(x_i)\cdot n - \varepsilon_i \nabla^2 u_0(x_i) n \cdot n\right) & \text{on } \partial B_{\varepsilon_i}(x_i). \end{cases} \tag{11.13}$$

where $\Xi_{\varepsilon_i} := B_{\varepsilon_i}(x_i) \bigcup \left(\mathbb{R}^2 \setminus \overline{B_{\varepsilon_i}(x_i)}\right)$. By fixing the notation

$$\rho_i = \frac{1-\gamma_i}{1+\gamma_i} \quad \text{and} \quad \gamma_{\varepsilon_i} = \gamma_{\varepsilon_i}(x) = \begin{cases} 1, & \text{if } x \in \mathbb{R}^2 \setminus B_{\varepsilon_i}(x_i) \\ \gamma_i, & \text{if } x \in B_{\varepsilon_i}(x_i), \end{cases} \tag{11.14}$$

the solutions of (11.13) in $\mathbb{R}^2 \setminus \overline{B_{\varepsilon_i}(x_i)}$ are given by

$$w_i(x/\varepsilon_i) = \varepsilon_i^2 g_i(x) + \varepsilon_i^4 h_i(x), \tag{11.15}$$

where

$$g_i(x) = \frac{\rho_i}{\|x-x_i\|^2} \nabla u_0(x_i) \cdot (x - x_i) \tag{11.16}$$

and

$$h_i(x) = \frac{\rho_i}{2\|x-x_i\|^4} \nabla^2 u_0(x_i)(x-x_i)^2. \tag{11.17}$$

The functions $w_i^j(x/\varepsilon_j)$, $j \neq i$, satisfy

$$\begin{cases} \operatorname{div}[q_\varepsilon(w_i^j)] = 0 & \text{in } \Xi_{\varepsilon_j}, \\ q_\varepsilon(w_i^j) = -\gamma_{\varepsilon_j} k \nabla w_i^j & \text{in } \mathbb{R}^2, \\ w_i^j \to 0 & \text{at } \infty \\ [\![w_i^j]\!] = 0 & \text{on } \partial B_{\varepsilon_j}(x_j), \\ [\![q_\varepsilon(w_i^j)]\!] \cdot n_j = k(1-\gamma_j)\varepsilon_i^2 A_i(x_j) \nabla u_0(x_i) \cdot n_j & \text{on } \partial B_{\varepsilon_j}(x_j). \end{cases} \tag{11.18}$$

The second order tensor $A_i(x)$ is defined by

$$A_i(x) = \frac{\rho_i}{\|x-x_i\|^2}\left[I - 2\frac{(x-x_i)\otimes(x-x_i)}{\|x-x_i\|^2}\right]. \tag{11.19}$$

Thus, the solutions of (11.18) in $\mathbb{R}^2 \setminus \overline{B_{\varepsilon_j}(x_j)}$ are given by

$$w_i^j(x/\varepsilon_j) = \varepsilon_i^2 \varepsilon_j^2 \theta_i^j(x), \tag{11.20}$$

where

$$\theta_i^j(x) = \frac{\rho_i \rho_j}{\|x-x_j\|^2} A_i(x_j) \nabla u_0(x_i) \cdot (x - x_j). \tag{11.21}$$

Since the boundary layers w_i introduce discrepancies on $\partial\Omega$, we construct $\tilde{u}_i$, $i = 1, \ldots, N$, such that

$$\begin{cases} \operatorname{div}[q(\tilde{u}_i)] = 0 & \text{in } \Omega, \\ q(\tilde{u}_i) = -k\nabla\tilde{u}_i & \text{in } \Omega, \\ q(\tilde{u}_i)\cdot n = -q(g_i)\cdot n & \text{on } \partial\Omega, \\ \int_{\Gamma_m}\tilde{u}_i = -\int_{\Gamma_m} g_i, \end{cases} \tag{11.22}$$

where g_i is given by (11.16).

The boundary layers $\tilde{w}_i(x/\varepsilon_i)$, $i = 1, \ldots, N$, satisfy

$$\begin{cases} \operatorname{div}[q_\varepsilon(\tilde{w}_i)] = 0 & \text{in } \Xi_{\varepsilon_i}, \\ q_\varepsilon(\tilde{w}_i) = -\gamma_{\varepsilon_i}k\nabla\tilde{w}_i & \text{in } \mathbb{R}^2, \\ \tilde{w}_i \to 0 & \text{at } \infty, \\ [\![\tilde{w}_i]\!] = 0 & \text{on } \partial B_{\varepsilon_i}(x_i), \\ [\![q_\varepsilon(\tilde{w}_i)]\!]\cdot n = k(1-\gamma_i)\varepsilon_i^2\left(\nabla\tilde{u}_i(x_i)\cdot n + \frac{1}{2}\nabla^3 u_0(x_i)n^3\right) & \text{on } \partial B_{\varepsilon_i}(x_i), \end{cases} \tag{11.23}$$

whose explicit solutions in $\mathbb{R}^2 \setminus \overline{B_{\varepsilon_i}(x_i)}$ are given by

$$\tilde{w}_i(x/\varepsilon) = \varepsilon_i^4\tilde{g}_i(x) + \varepsilon_i^6\tilde{h}_i(x), \tag{11.24}$$

where

$$\tilde{g}_i(x) = \frac{\rho_i}{\|x - x_i\|^2}\nabla\tilde{u}_i(x_i)\cdot(x - x_i) \tag{11.25}$$

and

$$\tilde{h}_i(x) = \frac{\rho_i}{2\|x - x_i\|^6}\nabla^3 u_0(x_i)(x - x_i)^3. \tag{11.26}$$

Now, we chose $\tilde{\tilde{u}}_i$ such that it compensates for the discrepancies of order $O(\varepsilon_i^4)$ left on the boundary $\partial\Omega$ by $\tilde{w}_i$ and w_i for $i = 1, \ldots, N$, namely,

$$\begin{cases} \operatorname{div}[q(\tilde{\tilde{u}}_i)] = 0 & \text{in } \Omega, \\ q(\tilde{\tilde{u}}_i) = -k\nabla\tilde{\tilde{u}}_i & \text{in } \Omega, \\ q(\tilde{\tilde{u}}_i)\cdot n = -q(h_i + \tilde{g}_i)\cdot n & \text{on } \partial\Omega, \\ \int_{\Gamma_m}\tilde{\tilde{u}}_i = -\int_{\Gamma_m} h_i + \tilde{g}_i. \end{cases} \tag{11.27}$$

The boundary layers w_i^j also produce discrepancies on $\partial\Omega$, which are compensated by u_i^j solution of the following boundary value problems for $i, j = 1, \ldots, N$, with $i \neq j$,

$$\begin{cases} \operatorname{div}[q(u_i^j)] = 0 & \text{in } \Omega, \\ q(u_i^j) = -k\nabla u_i^j & \text{in } \Omega, \\ q(u_i^j)\cdot n = -q(\theta_i^j)\cdot n, & \text{on } \partial\Omega, \\ \int_{\Gamma_m} u_i^j = -\int_{\Gamma_m}\theta_i^j, \end{cases} \tag{11.28}$$

where θ_i^j is given by (11.21).

Finally, the last term of the expansion (11.12), namely $\tilde{\tilde{u}}_\varepsilon$, has to compensate for all remainder terms, so that it is solution to the following boundary value problem

$$\begin{cases} \operatorname{div}[q_\varepsilon(\tilde{\tilde{u}}_\varepsilon)] = 0 & \text{in } \bigcup_{i=1}^N B_{\varepsilon_i}(x_i) \bigcup \left(\Omega \setminus \bigcup_{i=1}^N \overline{B_{\varepsilon_i}(x_i)}\right), \\ q_\varepsilon(\tilde{\tilde{u}}_\varepsilon) = -\gamma_\varepsilon k \nabla \tilde{\tilde{u}}_\varepsilon & \text{in } \Omega, \\ q_\varepsilon(\tilde{\tilde{u}}_\varepsilon) \cdot n = -\sum_{i=1}^N \varepsilon_i^6 q_\varepsilon(\tilde{h}_i) \cdot n & \text{on } \partial\Omega, \\ \int_{\Gamma_m} \tilde{\tilde{u}}_\varepsilon = -\sum_{i=1}^N \varepsilon_i^6 \int_{\Gamma_m} \tilde{h}_i & \\ [\![\tilde{\tilde{u}}_\varepsilon]\!] = 0 & \text{on } \bigcup_{i=1}^N \partial B_{\varepsilon_i}(x_i), \\ [\![q_\varepsilon(\tilde{\tilde{u}}_\varepsilon)]\!] \cdot n = \tilde{\tilde{g}}_\varepsilon & \text{on } \bigcup_{i=1}^N \partial B_{\varepsilon_i}(x_i), \end{cases} \tag{11.29}$$

where,

$$\tilde{\tilde{g}}_\varepsilon = \sum_{i=1}^N \sum_{j \neq i}^N \varepsilon_i^2 \varepsilon_j g_{ij}^a + \sum_{i=1}^N \varepsilon_i^3 g_i^b + \sum_{i=1}^N \varepsilon_i^4 (g_i^c + \partial_n h_i) + \sum_{i=1}^N \sum_{j \neq i}^N \varepsilon_i^2 \varepsilon_j^2 g_{ij}, \tag{11.30}$$

with

$$g_{ij}^a = (\nabla A_i(\xi_j) n_j) \nabla u_0(x_i) \cdot n_j, \tag{11.31}$$

$$g_i^b = k(1-\gamma_i) \frac{1}{3!} \nabla^4 u_0(\xi_i) n^4, \tag{11.32}$$

$$g_i^c = k(1-\gamma_i) \partial_n \tilde{\tilde{u}}_i, \tag{11.33}$$

$$g_{ij} = k \frac{1-\gamma_j}{\|x - x_j\|^2} A_i(x_j) \nabla u_0(x_i) \cdot (x - x_j), \tag{11.34}$$

for $\xi_j = \delta x + (1-\delta) x_j$, with $\delta \in (0,1)$ and $x \in B_{\varepsilon_j}(x_j)$.

Lemma 11.1 *Let $\tilde{\tilde{u}}_\varepsilon$ be solution to (11.29) or equivalently solution to the following variational problem: Find $\tilde{\tilde{u}}_\varepsilon \in \mathscr{U}_\varepsilon$, such that*

$$-\int_\Omega q_\varepsilon(\tilde{\tilde{u}}_\varepsilon) \cdot \nabla \eta + \int_{\partial B_\varepsilon} \tilde{\tilde{g}}_\varepsilon \eta - \int_{\partial\Omega} \sum_{i=1}^N \varepsilon_i^6 q_\varepsilon(\tilde{h}_i) \cdot n \eta = 0 \quad \forall \eta \in \mathscr{V}, \tag{11.35}$$

where the set $\mathscr{U}_\varepsilon$ and the space $\mathscr{V}$ are defined as

$$\mathscr{U}_\varepsilon := \left\{ \varphi \in H^1(\Omega) : \int_{\Gamma_m} \varphi = \sum_{i=1}^N \varepsilon_i^6 \rho_i \int_{\Gamma_m} \tilde{h}_i \right\} \tag{11.36}$$

and

$$\mathscr{V} := \left\{ \varphi \in H^1(\Omega) : \int_{\Gamma_m} \varphi = 0 \right\}. \tag{11.37}$$

Then, we have the estimate $\|\tilde{\tilde{u}}_\varepsilon\|_{H^1(\Omega)} = O(|\varepsilon|^5)$ *for the remainder, where* $|\varepsilon| := max\{\varepsilon_1, \ldots, \varepsilon_N\}$.

Proof By taking $\eta = \tilde{\tilde{u}}_\varepsilon - \varphi_\varepsilon$ in (11.35), with $\varphi_\varepsilon = \sum_{i=1}^{N} \varepsilon_i^6 \rho_i \tilde{h}_i$ on $\partial\Omega$, we have

$$-\int_\Omega q_\varepsilon(\tilde{\tilde{u}}_\varepsilon) \cdot \nabla \tilde{\tilde{u}}_\varepsilon + \int_\Omega q_\varepsilon(\tilde{\tilde{u}}_\varepsilon) \cdot \nabla \varphi_\varepsilon + \int_{\partial B_\varepsilon} \tilde{\tilde{g}}_\varepsilon \tilde{\tilde{u}}_\varepsilon - \int_{\partial B_\varepsilon} \tilde{\tilde{g}}_\varepsilon \varphi_\varepsilon$$
$$- \int_{\partial\Omega} \sum_{i=1}^{N} \varepsilon_i^6 q_\varepsilon(\tilde{h}_i) \cdot n \tilde{\tilde{u}}_\varepsilon + \int_{\partial\Omega} \sum_{i=1}^{N} \varepsilon_i^6 q_\varepsilon(\tilde{h}_i) \cdot n \varphi_\varepsilon = 0 \tag{11.38}$$

Integration by parts yields

$$-\int_\Omega q_\varepsilon(\tilde{\tilde{u}}_\varepsilon) \cdot \nabla \tilde{\tilde{u}}_\varepsilon - \int_\Omega \operatorname{div}[q_\varepsilon(\tilde{\tilde{u}}_\varepsilon)] \varphi_\varepsilon + \int_{\partial\Omega} q_\varepsilon(\tilde{\tilde{u}}_\varepsilon) \cdot n \varphi_\varepsilon + \int_{\partial B_\varepsilon} [\![q_\varepsilon(\tilde{\tilde{u}}_\varepsilon)]\!] \cdot n \varphi_\varepsilon$$
$$+ \int_{\partial B_\varepsilon} \tilde{\tilde{g}}_\varepsilon \tilde{\tilde{u}}_\varepsilon - \int_{\partial B_\varepsilon} \tilde{\tilde{g}}_\varepsilon \varphi_\varepsilon - \int_{\partial\Omega} \sum_{i=1}^{N} \varepsilon_i^6 q_\varepsilon(\tilde{h}_i) \cdot n \tilde{\tilde{u}}_\varepsilon + \int_{\partial\Omega} \sum_{i=1}^{N} \varepsilon_i^6 q_\varepsilon(\tilde{h}_i) \cdot n \varphi_\varepsilon = 0. \tag{11.39}$$

Using Eq. (11.29), we obtain the equality

$$-\int_\Omega q_\varepsilon(\tilde{\tilde{u}}_\varepsilon) \cdot \nabla \tilde{\tilde{u}}_\varepsilon = \int_{\partial\Omega} \sum_{i=1}^{N} \varepsilon_i^6 q_\varepsilon(\tilde{h}_i) \cdot n \tilde{\tilde{u}}_\varepsilon - \int_{\partial B_\varepsilon} \tilde{\tilde{g}}_\varepsilon \tilde{\tilde{u}}_\varepsilon \tag{11.40}$$

From the Cauchy–Schwarz inequality together with the trace theorem we have

$$-\int_\Omega q_\varepsilon(\tilde{\tilde{u}}_\varepsilon) \cdot \nabla \tilde{\tilde{u}}_\varepsilon \leq |\varepsilon|^6 \|\tilde{\tilde{u}}_\varepsilon\|_{L^2(\partial\Omega)} \left\| \sum_{i=1}^{N} q_\varepsilon(\tilde{h}_i) \right\|_{L^2(\partial\Omega)} + \|\tilde{\tilde{g}}_\varepsilon\|_{H^{\frac{1}{2}}(\partial B_\varepsilon)} \|\tilde{\tilde{u}}_\varepsilon\|_{H^{-\frac{1}{2}}(\partial B_\varepsilon)}$$
$$\leq C_1 |\varepsilon|^6 \|\tilde{\tilde{u}}_\varepsilon\|_{H^1(\Omega)} + C_2 \|\tilde{\tilde{g}}_\varepsilon\|_{H^1(B_\varepsilon)} \|\tilde{\tilde{u}}_\varepsilon\|_{L^2(B_\varepsilon)}. \tag{11.41}$$

Now, let us make use of the Hölder inequality together with the Sobolev embedding theorem for $1/p + 1/q$ and $q \geq 1$, to obtain

$$\|\tilde{\tilde{u}}_\varepsilon\|_{L^2(B_\varepsilon)} \leq C_3 |\varepsilon|^{1/q} \|\tilde{\tilde{u}}_\varepsilon\|_{L^{2p}(B_\varepsilon)} \leq |\varepsilon| C_4 \|\tilde{\tilde{u}}_\varepsilon\|_{H^1(\Omega)}, \tag{11.42}$$

where we have used the interior elliptic regularity of function $\tilde{\tilde{u}}_\varepsilon$. In addition, by using definition (11.30), we obtain

$$\|\tilde{\tilde{g}}_\varepsilon\|_{H^1(B_\varepsilon)} \leq C_5|\varepsilon|^4, \tag{11.43}$$

Therefore,

$$-\int_\Omega q_\varepsilon(\tilde{\tilde{u}}_\varepsilon) \cdot \nabla\tilde{\tilde{u}}_\varepsilon \leq C_6|\varepsilon|^5\|\tilde{\tilde{u}}_\varepsilon\|_{H^1(\Omega)}. \tag{11.44}$$

Finally, from the coercivity of the bilinear form on the left hand side of the above inequality we obtain

$$c\|\tilde{\tilde{u}}_\varepsilon\|^2_{H^1(\Omega)} \leq -\int_\Omega q_\varepsilon(\tilde{\tilde{u}}_\varepsilon) \cdot \nabla\tilde{\tilde{u}}_\varepsilon \leq C_6|\varepsilon|^5\|\tilde{\tilde{u}}_\varepsilon\|_{H^1(\Omega)}, \tag{11.45}$$

which leads to the result, namely $\|\tilde{\tilde{u}}_\varepsilon\|_{H^1(\Omega)} \leq C|\varepsilon|^5$, with constant $C = C_6/c$ independent of ε.

11.2.2 Asymptotic Expansion of the Shape Functional

From the ansätz for u_ε given by (11.12), we can obtain the asymptotic expansion of the shape functional $\mathscr{J}_\varepsilon(u_\varepsilon)$ defined through (11.9) with respect to ε. In fact, the shape functional is defined on the boundary Γ_m. Then, let us evaluate the expansion for u_ε on the boundary $\partial\Omega$ to obtain $u_{\varepsilon|\partial\Omega} = (u_0 + \varphi_\varepsilon)_{|\partial\Omega}$, where φ_ε is such that

$$\varphi_\varepsilon = \sum_{i=1}^{N}\left(\varepsilon_i^2(g_i + \tilde{u}_i) + \varepsilon_i^4(h_i + \tilde{g}_i + \tilde{\tilde{u}}_i) + \varepsilon_i^6\tilde{h}_i\right) + \sum_{i=1}^{N}\sum_{j\neq i}^{N}\left(\varepsilon_i^2\varepsilon_j^2(\theta_i^j + u_i^j)\right) + \tilde{\tilde{u}}_\varepsilon. \tag{11.46}$$

Therefore,

$$\mathscr{J}_\varepsilon(u_\varepsilon) = \int_{\Gamma_m}(u_0 + \varphi_\varepsilon - U)^2 = \mathscr{J}_0(u_0) + 2\int_{\Gamma_m}(u_0 - U)\varphi_\varepsilon + \int_{\Gamma_m}\varphi_\varepsilon^2. \tag{11.47}$$

Let us now collect the terms on the right-hand side of (11.47) in power of ε. The first one is independent of ε. It is actually the original shape functional. In view of (11.46), the second term can be written as

$$\begin{aligned}\int_{\Gamma_m}(u_0 - U)\varphi_\varepsilon = &\sum_{i=1}^{N}\left(\varepsilon_i^2\int_{\Gamma_m}(u_0 - U)(g_i + \tilde{u}_i) + \varepsilon_i^4\int_{\Gamma_m}(u_0 - U)(h_i + \tilde{g}_i + \tilde{\tilde{u}}_i)\right)\\ &+ \sum_{i=1}^{N}\sum_{j\neq i}^{N}\varepsilon_i^2\varepsilon_j^2\int_{\Gamma_m}(u_0 - U)(\theta_i^j + u_i^j) + \sum_{\ell=1}^{2}\mathscr{E}_\ell(\varepsilon),\end{aligned} \tag{11.48}$$

where, from the Cauchy–Scharwz inequality together with Lemma 11.1, we have

$$\mathscr{E}_1(\varepsilon) = \sum_{i=1}^{N} \varepsilon_i^6 \int_{\Gamma_m} (u_0 - U)\tilde{h}_i = O(|\varepsilon|^6), \tag{11.49}$$

$$\mathscr{E}_2(\varepsilon) = \int_{\Gamma_m} (u_0 - U)\tilde{\tilde{u}}_\varepsilon = O(|\varepsilon|^5), \tag{11.50}$$

The last term on the right-hand side of (11.47) can be expanded as follows

$$\int_{\Gamma_m} \varphi_\varepsilon^2 = \int_{\Gamma_m} \left(\sum_{i=1}^{N} \varepsilon_i^2 (g_i + \tilde{u}_i) \right)^2 + \sum_{\ell=3}^{15} \mathscr{E}_\ell(\varepsilon), \tag{11.51}$$

with

$$\mathscr{E}_3(\varepsilon) = 2 \int_{\Gamma_m} \sum_{i=1}^{N} \varepsilon_i^2 (g_i + \tilde{u}_i) \sum_{i=1}^{N} \varepsilon_i^4 (h_i + \tilde{g}_i + \tilde{\tilde{u}}_i) = O(|\varepsilon|^6), \tag{11.52}$$

$$\mathscr{E}_4(\varepsilon) = 2 \int_{\Gamma_m} \left(\sum_{i=1}^{N} \varepsilon_i^4 (h_i + \tilde{g}_i + \tilde{\tilde{u}}_i) \right)^2 = O(|\varepsilon|^8), \tag{11.53}$$

$$\mathscr{E}_5(\varepsilon) = 2 \int_{\Gamma_m} \sum_{i=1}^{N} \sum_{j \neq i}^{N} \varepsilon_i^2 \varepsilon_j^2 (\theta_i^j + u_i^j) \left(\sum_{i=1}^{N} \varepsilon_i^2 (g_i + \tilde{u}_i) \right) = O(|\varepsilon|^6), \tag{11.54}$$

$$\mathscr{E}_6(\varepsilon) = 2 \int_{\Gamma_m} \sum_{i=1}^{N} \sum_{j \neq i}^{N} \varepsilon_i^2 \varepsilon_j^2 (\theta_i^j + u_i^j) \left(\sum_{i=1}^{N} \varepsilon_i^4 (h_i + \tilde{g}_i + \tilde{\tilde{u}}_i) \right) = O(|\varepsilon|^8), \tag{11.55}$$

$$\mathscr{E}_7(\varepsilon) = \int_{\Gamma_m} \left(\sum_{i=1}^{N} \sum_{j \neq i}^{N} \varepsilon_i^2 \varepsilon_j^2 (\theta_i^j + u_i^j) \right)^2 = O(|\varepsilon|^8), \tag{11.56}$$

$$\mathscr{E}_8(\varepsilon) = \int_{\Gamma_m} \sum_{i=1}^{N} \varepsilon_i^6 \tilde{h}_i \sum_{i=1}^{N} \varepsilon_i^2 (g_i + \tilde{u}_i) = O(|\varepsilon|^8), \tag{11.57}$$

$$\mathscr{E}_9(\varepsilon) = 2 \int_{\Gamma_m} \sum_{i=1}^{N} \varepsilon_i^6 \tilde{h}_i \sum_{i=1}^{N} \varepsilon_i^4 (h_i + \tilde{g}_i + \tilde{\tilde{u}}_i) = O(|\varepsilon|^{10}), \tag{11.58}$$

$$\mathscr{E}_{10}(\varepsilon) = \int_{\Gamma_m} \left(\sum_{i=1}^{N} \varepsilon_i^6 \tilde{h}_i \right)^2 = O(|\varepsilon|^{12}), \tag{11.59}$$

$$\mathscr{E}_{11}(\varepsilon) = \int_{\Gamma_m} \sum_{i=1}^{N} \varepsilon_i^6 \tilde{h}_i \tilde{\tilde{u}}_\varepsilon = O(|\varepsilon|^{11}), \tag{11.60}$$

$$\mathscr{E}_{12}(\varepsilon) = 2 \int_{\Gamma_m} \tilde{\tilde{u}}_\varepsilon \sum_{i=1}^{N} \varepsilon_i^2 (g_i + \tilde{u}_i) = O(|\varepsilon|^7), \tag{11.61}$$

$$\mathscr{E}_{13}(\varepsilon) = 2 \int_{\Gamma_m} \tilde{\tilde{u}}_\varepsilon \sum_{i=1}^{N} \varepsilon_i^4 (h_i + \tilde{g}_i + \tilde{\tilde{u}}_i) = O(|\varepsilon|^9), \tag{11.62}$$

$$\mathscr{E}_{14}(\varepsilon) = \int_{\Gamma_m} \tilde{\tilde{u}}_\varepsilon \sum_{i=1}^{N} \varepsilon_i^6 \tilde{h}_i = O(|\varepsilon|^{11}), \tag{11.63}$$

$$\mathscr{E}_{15}(\varepsilon) = \int_{\Gamma_m} \tilde{\tilde{u}}_\varepsilon^2 = O(|\varepsilon|^{10}), \tag{11.64}$$

where we have used again the Cauchy–Scharwz inequality together with Lemma 11.1.

Finally, after replacing (11.48) and (11.51) into (11.47), we obtain the following asymptotic expansion for the topologically perturbed shape functional $\mathscr{J}_\varepsilon(u_\varepsilon)$

$$\begin{aligned} \mathscr{J}_\varepsilon(u_\varepsilon) &= \mathscr{J}_0(u_0) + 2\sum_{i=1}^{N} \left(\varepsilon_i^2 \int_{\Gamma_m} (u_0 - U)(g_i + \tilde{u}_i) + \varepsilon_i^4 \int_{\Gamma_m} (u_0 - U)(h_i + \tilde{g}_i + \tilde{\tilde{u}}_i) \right) \\ &+ 2\sum_{i=1}^{N} \sum_{j \neq i}^{N} \varepsilon_i^2 \varepsilon_j^2 \int_{\Gamma_m} (u_0 - U)(\theta_i^j + u_i^j) + \int_{\Gamma_m} \left(\sum_{i=1}^{N} \varepsilon_i^2 (g_i + \tilde{u}_i) \right)^2 + \mathscr{E}(\varepsilon), \end{aligned} \tag{11.65}$$

with

$$\mathscr{E}(\varepsilon) = \sum_{\ell=1}^{15} \mathscr{E}_\ell(\varepsilon) = O(|\varepsilon|^5). \tag{11.66}$$

11.2.3 Introduction of Adjoint States

Note that to evaluate expansion (11.65), we have to solve the problems associated with the non-local terms $\tilde{u}_i$ and $\tilde{\tilde{u}}_i$ for each point x_i, and u_i^j for each pair of points x_i and x_j. However, thanks to the representation we have found, the non-local terms which appears in the first, second and third integrals in (11.65) can be replaced by just one adjoint state independent of the points x_i and x_j. On the other hand, the term $\tilde{u}_i$ also appears in the last integral of expansion (11.65) in a quadratic form, so that $\tilde{u}_i$ has to be computed. Therefore, the adjoint state will be used to replace only the terms involving $\tilde{\tilde{u}}_i$ and u_i^j. In fact, expansion (11.65) can be rewritten as

$$\begin{aligned}\mathscr{J}_\varepsilon(u_\varepsilon) = \mathscr{J}_0(u_0) + 2\sum_{i=1}^{N}\left(\varepsilon_i^2 \int_{\Gamma_m}(u_0 - U)(g_i + \tilde{u}_i) + \varepsilon_i^4 \int_{\Gamma_m}(u_0 - U)(h_i + \tilde{g}_i)\right) \\ + 2\sum_{i=1}^{N}\left(\varepsilon_i^4 \int_{\Gamma_m}(u_0 - U)\tilde{\tilde{u}}_i + \sum_{j\neq i}^{N}\varepsilon_i^2\varepsilon_j^2 \int_{\Gamma_m}(u_0 - U)u_i^j\right) \\ + 2\sum_{i=1}^{N}\sum_{j\neq i}^{N}\left(\varepsilon_i^2\varepsilon_j^2 \int_{\Gamma_m}(u_0 - U)\theta_i^j\right) + \int_{\Gamma_m}\left(\sum_{i=1}^{N}\varepsilon_i^2(g_i + \tilde{u}_i)\right)^2 + \mathscr{E}(\varepsilon).\end{aligned} \tag{11.67}$$

Let us introduce an adjoint state solution of the following variational problem: Find $v \in \mathscr{V}$, such that

$$\int_\Omega q(v)\cdot\nabla\eta = 2\int_{\Gamma_m}(u_0 - U)\,\eta, \quad \forall \eta \in \mathscr{V}, \tag{11.68}$$

where the space $\mathscr{V}$ is given by (11.37). The associated strong form of (11.68) is written as

$$\begin{cases} \operatorname{div}[q(v)] = 0 & \text{in } \Omega, \\ q(v) = -k\nabla v, & \\ q(v)\cdot n = 2(u_0 - U) & \text{on } \Gamma_m, \\ q(v)\cdot n = 0 & \text{on } \partial\Omega \setminus \Gamma_m, \\ \int_{\Gamma_m} v = 0. & \end{cases} \tag{11.69}$$

The weak form of (11.27) reads: Find $\tilde{\tilde{u}}_i \in \mathscr{U}_i$, such that

$$\int_\Omega q(\tilde{\tilde{u}}_i)\cdot\nabla\eta + \int_{\partial\Omega} q(h_i + \tilde{g}_i)\cdot n\,\eta = 0, \quad \forall \eta \in \mathscr{V}, \tag{11.70}$$

where the space $\mathscr{V}$ is given by (11.37) and the set $\mathscr{U}_i$ is defined as

$$\mathscr{U}_i := \left\{\varphi \in H^1(\Omega) : \int_{\Gamma_m}\varphi = -\int_{\Gamma_m} h_i + \tilde{g}_i\right\}. \tag{11.71}$$

By setting $\eta = \tilde{\tilde{u}}_i + \varphi_i$ as test function in (11.68), with $\varphi_i = h_i + \tilde{g}_i$ on $\partial\Omega$, we obtain the equality

$$\int_{\Omega} q(v) \cdot \nabla(\tilde{\tilde{u}}_i + \varphi_i) = 2\int_{\Gamma_m} (u_0 - U)(\tilde{\tilde{u}}_i + \varphi_i). \tag{11.72}$$

Integration by parts yields

$$\begin{aligned} \int_{\Omega} q(v) \cdot \nabla\tilde{\tilde{u}}_i &= 2\int_{\Gamma_m} (u_0 - U)\tilde{\tilde{u}}_i + 2\int_{\Gamma_m} (u_0 - U)\varphi_i - \int_{\partial\Omega} q(v) \cdot n\varphi_i \\ &= 2\int_{\Gamma_m} (u_0 - U)\tilde{\tilde{u}}_i, \end{aligned} \tag{11.73}$$

since v solves (11.69). Now, let us set $\eta = v$ as test function in (11.70) to obtain the equality

$$\int_{\Omega} q(\tilde{\tilde{u}}_i) \cdot \nabla v = -\int_{\partial\Omega} q(h_i + \tilde{g}_i) \cdot n\, v. \tag{11.74}$$

After comparing the obtained results we have the following important identity

$$2\int_{\Gamma_m} (u_0 - U)\tilde{\tilde{u}}_i = -\int_{\partial\Omega} q(h_i + \tilde{g}_i) \cdot n\, v. \tag{11.75}$$

In addition, the weak form of (11.28) can be written as: Find $u_i^j \in \mathscr{U}_i^j$, such that

$$\int_{\Omega} q(u_i^j) \cdot \nabla\eta + \int_{\partial\Omega} q(\theta_i^j) \cdot n\, \eta = 0, \quad \forall \eta \in \mathscr{V}, \tag{11.76}$$

with the space $\mathscr{V}$ given by (11.37) and the set $\mathscr{U}_i^j$ given by

$$\mathscr{U}_i^j := \left\{ \varphi \in H^1(\Omega);\ \int_{\Gamma_m} \varphi = -\int_{\Gamma_m} \theta_i^j \right\}. \tag{11.77}$$

After setting $\eta = u_i^j + \varphi_i^j$ as test function in (11.68), with $\varphi_i^j = \theta_i^j$ on $\partial\Omega$, there is

$$\int_{\Omega} q(v) \cdot \nabla(u_i^j + \varphi_i^j) = 2\int_{\Gamma_m} (u_0 - U)(u_i^j + \varphi_i^j). \tag{11.78}$$

From integration by parts we obtain

$$\begin{aligned} \int_{\Omega} q(v) \cdot \nabla u_i^j &= 2\int_{\Gamma_m} (u_0 - U)u_i^j + 2\int_{\Gamma_m} (u_0 - U)\varphi_i^j - \int_{\partial\Omega} q(v) \cdot n\varphi_i \\ &= 2\int_{\Gamma_m} (u_0 - U)u_i^j, \end{aligned} \tag{11.79}$$

where we have used (11.69). By comparing the last two results, the following important equality holds true

$$2\int_{\Gamma_m}(u_0 - U)u_i^j = -\int_{\partial\Omega} q(\theta_i^j)\cdot n\,v. \tag{11.80}$$

Finally, we can respectively replace the third and fourth integrals in (11.67) by the obtained equalities (11.75) and (11.80), namely

$$\begin{aligned}\mathscr{J}(u_\varepsilon) = \mathscr{J}(u_0) + 2\sum_{i=1}^N \varepsilon_i^2 \int_{\Gamma_m}(u_0 - U)(g_i + \tilde{u}_i)\\ + \sum_{i=1}^N \varepsilon_i^4\left(2\int_{\Gamma_m}(u_0 - U)(h_i + \tilde{g}_i) - \int_{\partial\Omega} q(h_i + \tilde{g}_i)\cdot n\,v\right)\\ + \sum_{i=1}^N\sum_{j\neq i}^N \varepsilon_i^2\varepsilon_j^2\left(2\int_{\Gamma_m}(u_0 - U)\theta_i^j - \int_{\partial\Omega} q(\theta_i^j)\cdot nv\right) + \int_{\Gamma_m}\left(\sum_{i=1}^N \varepsilon_i^2(g_i + \tilde{u}_i)\right)^2 + \mathscr{E}(\varepsilon).\end{aligned} \tag{11.81}$$

Therefore, without any approximation, the integrals in (11.67) involving the non-local terms $\tilde{u}_i$ and u_i^j have been replaced by just one adjoint state v, solution of (11.69), which does not depend on the points x_i and x_j.

11.3 A Noniterative Reconstruction Algorithm

In this section we present the resulting non-iterative reconstruction algorithm based on the expansion (11.81). The topological asymptotic expansion of the shape functional $\mathscr{J}(u_\varepsilon)$ given by (11.81) can be rewritten in the following compact form

$$\mathscr{J}(u_\varepsilon) = \mathscr{J}(u_0) + d(\xi)\cdot\alpha + \frac{1}{2}H(\xi)\alpha\cdot\alpha + \mathscr{E}(\varepsilon), \tag{11.82}$$

where $d(\xi)$ and $H(\xi)$ are the first and second order topological derivatives, respectively. In addition, $\alpha = (\varepsilon_1^2, \ldots, \varepsilon_N^2)$ and $\mathscr{E}(\varepsilon)$ is the remainder.

The vector $d(\xi)$ and the matrix $H(\xi)$ are defined as

$$d(\xi) := \begin{pmatrix} d_1 \\ \vdots \\ d_N \end{pmatrix} \quad \text{and} \quad H(\xi) := \begin{pmatrix} h_{11} & h_{12} & \cdots & h_{1N} \\ h_{21} & h_{22} & \cdots & h_{2N} \\ \vdots & \vdots & \ddots & \vdots \\ h_{N1} & h_{N2} & \cdots & h_{NN}, \end{pmatrix} \tag{11.83}$$

where each component d_i of the topological derivative vector $d(\xi)$ is given by

$$d_i := b(i) = -2\int_{\Gamma_m}(u_0 - U)(g_i + \tilde{u}_i) \tag{11.84}$$

while the entries $h_{ij} := A(i, j)$ of the topological Hessian matrix $H(\xi)$ are defined as

$$A(i,i) = 4\int_{\Gamma_m} (u_0 - U)(h_i + \tilde{g}_i) - 2\int_{\partial\Omega} (q(h_i + \tilde{g}_i)) \cdot n\, v + 2\int_{\Gamma_m} (g_i + \tilde{u}_i)^2, \tag{11.85}$$

and, for $i \neq j$,

$$A(i,j) = 2\int_{\Gamma_m} (u_0 - U)(\theta_i^j + \theta_j^i) - \int_{\partial\Omega} q(\theta_i^j + \theta_j^i) \cdot n\, v + 2\int_{\Gamma_m} (g_i + \tilde{u}_i)(g_j + \tilde{u}_j). \tag{11.86}$$

In addition, the functions $g_i(x)$, $h_i(x)$, $\tilde{g}_i(x)$ and $\theta_i^j(x)$ are respectively given by (11.16), (11.17), (11.25) and (11.21). Finally, the auxiliary function $\tilde{u}_i$ solves (11.22) and v is solution to the adjoint equation (11.69).

Note that the expression on the right-hand side of (11.81) depends explicitly on the number N of anomalies, their positions xi and sizes α. Thus, let us now introduce the quantity

$$\Psi(\xi, \alpha, N) = d(\xi) \cdot \alpha + \frac{1}{2} H(\xi)\alpha \cdot \alpha. \tag{11.87}$$

After minimizing (11.87) with respect to α we obtain the following linear system

$$\alpha = -(H(\xi))^{-1} d(\xi). \tag{11.88}$$

Let us replace $\alpha = \alpha(\xi)$ solution of (11.88) in (11.87), to obtain

$$\Psi(\xi, \alpha(\xi), N) = -\frac{1}{2} d(\xi) \cdot \alpha(\xi). \tag{11.89}$$

Therefore, the pair of vectors $(\xi^\star, \alpha^\star)$ which minimizes (11.87) is given by

$$\xi^\star := \arg\min_{\xi \in X} \left\{ -\frac{1}{2} d(\xi) \cdot \alpha(\xi) \right\} \quad \text{and} \quad \alpha^\star := \alpha(\xi^\star), \tag{11.90}$$

where X is the set of admissible locations of the inclusions. From these elements the Algorithm 2 presented in Chap. 10 can be used.

11.4 Numerical Experiments

Let us consider a disk of unitary radius with center at the origin, namely $\Omega = B_1(0)$. Its electrical conductivity is assumed to be uniform and given by $k = 1$. The boundary of the disk $\partial\Omega$ is subdivided into 16 disjoint pieces representing the electrodes. One pair of such electrodes is selected for injecting and draining the electrical current. Therefore, the excitation Q is given by a pair $Q_{in} = 1$ of injection and $Q_{out} = -1$ of draining. The remainder part of the boundary $\partial\Omega$ remains insulated. The associated potential U is measured on a part Γ_m of the boundary of the disk $\partial\Omega$. For more than one measurement, this procedure is repeated by changing the selected pair of injection and draining electrodes. From these information we are going to reconstruct an unknown number of anomalies with contrast $\gamma_i = 2$, for $i = 1, \ldots, N^*$. See sketch in Fig. 11.2. The thick lines represent the electrodes, where measurements are taken.

The auxiliary boundary value problems are solved using a finite element mesh with 32768 elements and 16641 nodes. From these solutions the sensitivities can be numerically evaluated at any point of the mesh. However, because of the high complexity of Algorithm 2 [181], a sub-mesh is defined over the finite element mesh where the combinatorial search is performed, leading to the optimal solution $(\alpha^\star, \xi^\star)$ defined in the sub-mesh.

We are interested in investigating the robustness of the method with respect to noisy data, so that the electrical conductivity k^* is corrupted with White Gaussian Noise (WGN) of zero mean and standard deviation μ. Therefore, k^* is replaced by $k^*_\mu = k^*(1 + \mu\,\nu)$, where ν is a function assuming random values in the interval $(-1, 1)$ and μ corresponds to the noise level.

The target consists of three ball-shaped anomalies, which is corrupted with different levels of noise $\mu = 10\%$, $\mu = 15\%$ and $\mu = 20\%$, as shown in Fig. 11.3, left. The associated reconstructions obtained with $M = 64$ boundary measurements are shown in Fig. 11.3, right.

11.5 Final Remarks

In this chapter a new reconstruction method for a class of electrical impedance tomography problems has been proposed. It relies on the topological derivatives concept. The basic idea consists in rewrite the inverse problem as a topology optimization problem, where a shape functional measuring the misfit between the boundary measurements and the electrical potentials obtained from the model has been minimized with respect to a set of ball-shaped anomalies of different sizes. The adjoint method has been evoked a posteriori, after obtaining the associated sensitivities, allowing us to derive a simpler representation for the resulting expansion, which has been truncated up to the second order term, leading to a quadratic and strictly convex form with respect to the volume of the inclusions. Therefore, the truncated expansion has been used to devise a novel non-iterative reconstruction algorithm based on a simple optimization step. As a result, the reconstruction process has become very robust with respect to noisy data and also independent of any initial guess.

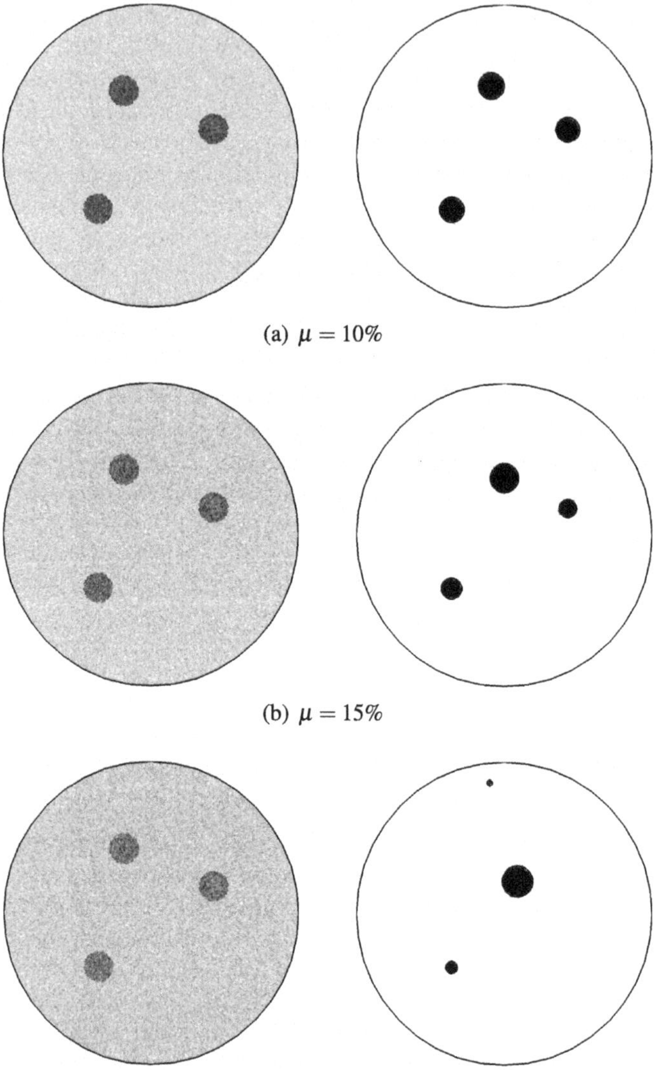

(a) $\mu = 10\%$

(b) $\mu = 15\%$

(c) $\mu = 20\%$

Fig. 11.3 Target corrupted with different level of White Gaussian Noise (left) and obtained results with $M = 64$ partial boundary measurements (right)

References

1. N. Aage, E. Andreassen, B.S. Lazarov, O. Sigmund, Giga-voxel computational morphogenesis for structural design. Nature **550**, 84–86 (2017)
2. G.S. Alberti, H. Ammari, B. Jin, J.-K. Seo, W. Zhang, The linearized inverse problem in multifrequency electrical impedance tomography. SIAM J. Imaging Sci. **9**(4), 1525–1551 (2016)
3. G. Allaire, *Shape Optimization by the Homogenization Method*, vol. 146. Applied Mathematical Sciences (Springer, New York, 2002)
4. G. Allaire, *Conception Optimale de Structures*, vol. 58. Mathématiques et Applications (Springer, Berlin, 2007)
5. G. Allaire, E. Bonnetier, G. Francfort, F. Jouve, Shape optimization by the homogenization method. Numer. Math. **76**(1), 27–68 (1997)
6. G. Allaire, F. de Gournay, F. Jouve, A.M. Toader, Structural optimization using topological and shape sensitivity via a level set method. Control Cybern. **34**(1), 59–80 (2005)
7. G. Allaire, F. Jouve, H. Maillot, Minimum stress optimal design with the level-set method. Eng. Anal. Bound. Elem. **32**(11), 909–918 (2008)
8. G. Allaire, F. Jouve, N. Van Goethem, Damage and fracture evolution in brittle materials by shape optimization methods. J. Comput. Phys. **230**(12), 5010–5044 (2011)
9. A.A.S. Amad, A.F.D. Loula, A.A. Novotny, A new method for topology design of electromagnetic antennas in hyperthermia therapy. Appl. Math. Model. **42**, 209–222 (2017)
10. R.C.R. Amigo, S.M. Giusti, A.A. Novotny, E.C.N. Silva, J. Sokolowski, Optimum design of flextensional piezoelectric actuators into two spatial dimensions. SIAM J. Control Optim. **52**(2), 760–789 (2016)
11. H. Ammari, E. Bretin, J. Garnier, H. Kang, H. Lee, A. Wahab, *Mathematical Methods in Elasticity Imaging*. Princeton Series in Applied Mathematics (Princeton University Press, Princeton, 2015)
12. H. Ammari, J. Garnier, W. Jing, H. Kang, M. Lim, K. Sølna, H. Wang, *Mathematical and Statistical Methods for Multistatic Imaging*, vol. 2098 (Springer, Berlin, 2013)
13. H. Ammari, J. Garnier, V. Jugnon, H. Kang, Stability and resolution analysis for a topological derivative based imaging functional. SIAM J. Control. Optim. **50**(1), 48–76 (2012)
14. H. Ammari, H. Kang, High-order terms in the asymptotic expansions of the steady-state voltage potentials in the presence of inhomogeneities of small diameter. SIAM J. Math. Anal. **34**(5), 1152–1166 (2003)
15. H. Ammari, H. Kang, *Reconstruction of Small Inhomogeneities from Boundary Measurements*. Lectures Notes in Mathematics, vol. 1846 (Springer, Berlin, 2004)
16. H. Ammari, H. Kang, *Polarization and Moment Tensors with Applications to Inverse Problems and Effective Medium Theory*. Applied Mathematical Sciences, vol. 162 (Springer, New York, 2007)

A. A. Novotny et al., *Applications of the Topological Derivative Method*, Studies in Systems, Decision and Control 188,
https://doi.org/10.1007/978-3-030-05432-8

17. H. Ammari, H. Kang, H. Lee, J. Lim, Boundary perturbations due to the presence of small linear cracks in an elastic body. J. Elast. **113**, 75–91 (2013)
18. H. Ammari, H. Kang, G. Nakamura, K. Tanuma, Complete asymptotic expansions of solutions of the system of elastostatics in the presence of inhomogeneities of small diameter. J. Elast. **67**, 97–129 (2002)
19. H. Ammari, A. Khelifi, Electromagnetic scattering by small dielectric inhomogeneities. J. Math. Pures Appl. **82**, 749–842 (2003)
20. H. Ammari, G. Uhlmann, Reconstruction of the potential from partial Cauchy data for the Schrödinger equation. Indiana Univ. Math. J. **53**(1), 169–183 (2004)
21. S. Amstutz, The topological asymptotic for the Navier-Stokes equations. ESAIM: Control Optim. Calc. Var. **11**(3), 401–425 (2005)
22. S. Amstutz, Sensitivity analysis with respect to a local perturbation of the material property. Asymptot. Anal. **49**(1–2), 87–108 (2006)
23. S. Amstutz, A penalty method for topology optimization subject to a pointwise state constraint. ESAIM: Control Optim. Calc. Var. **16**(3), 523–544 (2010)
24. S. Amstutz, Analysis of a level set method for topology optimization. Optim. Methods Softw. **26**(4–5), 555–573 (2011)
25. S. Amstutz, H. Andrä, A new algorithm for topology optimization using a level-set method. J. Comput. Phys. **216**(2), 573–588 (2006)
26. S. Amstutz, A. Bonnafé, Topological derivatives for a class of quasilinear elliptic equations. J. Math. Pures Appl. **107**, 367–408 (2017)
27. S. Amstutz, M.A. Ciligot-Travain, A notion of compliance robustness in topology optimization. ESAIM: Control Optim. Calc. Var. **22**(1), 64–87 (2016)
28. S. Amstutz, N. Dominguez, Topological sensitivity analysis in the context of ultrasonic non-destructive testing. Eng. Anal. Bound. Elem. **32**(11), 936–947 (2008)
29. S. Amstutz, S.M. Giusti, A.A. Novotny, E.A. de Souza Neto, Topological derivative for multi-scale linear elasticity models applied to the synthesis of microstructures. Int. J. Numer. Methods Eng. **84**, 733–756 (2010)
30. S. Amstutz, I. Horchani, M. Masmoudi, Crack detection by the topological gradient method. Control Cybern. **34**(1), 81–101 (2005)
31. S. Amstutz, A.A. Novotny, Topological optimization of structures subject to von Mises stress constraints. Struct. Multidiscip. Optim. **41**(3), 407–420 (2010)
32. S. Amstutz, A.A. Novotny, Topological asymptotic analysis of the Kirchhoff plate bending problem. ESAIM: Control Optim. Calc. Var. **17**(3), 705–721 (2011)
33. S. Amstutz, A.A. Novotny, E.A. de Souza Neto, Topological derivative-based topology optimization of structures subject to Drucker-Prager stress constraints. Comput. Methods Appl. Mech. Eng. **233–236**, 123–136 (2012)
34. S. Amstutz, A.A. Novotny, N. Van Goethem, Minimal partitions and image classification using a gradient-free perimeter approximation. Inverse Probl. Imaging **8**(2), 361–387 (2014)
35. S. Amstutz, A.A. Novotny, N. Van Goethem, Topological sensitivity analysis for elliptic differential operators of order $2m$. J. Differ. Equs. **256**, 1735–1770 (2014)
36. I.I. Argatov, Asymptotic models for the topological sensitivity versus the topological derivative. Open Appl. Math. J. **2**, 20–25 (2008)
37. I.I. Argatov, AFM-based indentation stiffness tomography - an asymptotic model. J. Mech. Phys. Solids **70**, 190–199 (2014)
38. I.I. Argatov, J. Sokolowski, Asymptotics of the energy functional of the Signorini problem under a small singular perturbation of the domain. Comput. Math. Math. Phys. **43**, 710–724 (2003)
39. I.I. Argatov, J. Sokołowski, On asymptotic behavior of the energy functional for the Signorini problem under small singular perturbation of the domain. Comput. Math. Math. Phys. **43**, 742–756 (2003)
40. D. Auroux, L. Jaafar Belaid, M. Masmoudi, A topological asymptotic analysis for the regularized greylevel image classification problem. ESAIM Math. Model. Numer. Anal. **41**, 607–625 (2007)

41. E. Bednarczuk, M. Pierre, E. Rouy, J. Sokolowski, Tangent sets in some functional spaces. Nonlinear Anal. **42**(5, Ser. A: Theory Methods), 871–886 (2000)
42. L.J. Belaid, M. Jaoua, M. Masmoudi, L. Siala, Application of the topological gradient to image restoration and edge detection. Eng. Anal. Bound. Elem. **32**(11), 891–899 (2008)
43. C. Bellis, M. Bonnet, F. Cakoni, Acoustic inverse scattering using topological derivative of far-field measurements-based L^2 cost functionals. Inverse Probl. **29**, 075012 (2013)
44. S. Belov, N. Fujii, Symmetry and sufficient condition of optimality in a domain optimization problem. Control Cybern. **26**(1), 45–56 (1997)
45. A. Ben Abda, M. Hassine, M. Jaoua, M. Masmoudi, Topological sensitivity analysis for the location of small cavities in stokes flow. SIAM J. Control Optim. **48**, 2871–2900 (2009)
46. M.P. Bendsøe, *Optimization of Structural Topology, Shape, and Material* (Springer, Berlin, 1995)
47. M.P. Bendsøe, O. Sigmund, *Topology Optimization. Theory, Methods and Applications* (Springer, Berlin, 2003)
48. A. Bensoussan, J.-L. Lions, *Applications of Variational Inequalities in Stochastic Control*. Volume 12 of Studies in Mathematics and its Applications (North-Holland Publishing Co., Amsterdam-New York, 1982) (Translated from the French)
49. P. Beremlijski, J. Haslinger, J.V. Outrata, R. Pathó, Shape optimization in contact problems with coulomb friction and a solution-dependent friction coefficient. SIAM J. Control Optim. **52**(5), 3371–3400 (2014)
50. E. Beretta, E. Bonnetier, E. Francini, A.L. Mazzucato, Small volume asymptotics for anisotropic elastic inclusions. Inverse Probl. Imaging **6**(1), 1–23 (2012)
51. E. Beretta, Y. Capdeboscq, F. Gournay, E. Francini, Thin cylindrical conductivity inclusions in a three-dimensional domain: a polarization tensor and unique determination from boundary data. Inverse Probl. **25**(6), 065004 (2009)
52. D. Bojczuk, Z. Mróz, Topological sensitivity derivative and finite topology modifications: application to optimization of plates in bending. Struct. Multidiscip. Optim. **39**(1), 1–15 (2009)
53. V. Bonnaillie-Noël, M. Dambrine, Interactions between moderately close circular inclusions: the Dirichlet-Laplace equation in the plane. Asymptot. Anal. **84**(3–4), 197–227 (2013)
54. M. Bonnet, Topological sensitivity for 3d elastodynamic and acoustic inverse scattering in the time domain. Comput. Methods Appl. Mech. Eng. **195**(37–40), 5239–5254 (2006)
55. M. Bonnet, Higher-order topological sensitivity for 2-D potential problems. Int. J. Solids Struct. **46**(11–12), 2275–2292 (2009)
56. M. Bonnet, R. Cornaggia, Higher order topological derivatives for three-dimensional anisotropic elasticity. ESAIM Control Optim. Calc. Var. **51**(6), 2069–2092 (2017)
57. G. Bouchitté, I. Fragalà, I. Lucardesi, Shape derivatives for minima of integral functionals. Math. Program. **148(1–2**, Ser. B), 111–142 (2014)
58. G. Bouchitté, I. Fragalà, I. Lucardesi, A variational method for second order shape derivatives. SIAM J. Control Optim. **54**(2), 1056–1084 (2016)
59. M. Bruggi, P. Duysinx, Topology optimization for minimum weight with compliance and stress constraints. Struct. Multidiscip. Optim. **46**(3), 369–384 (2012)
60. M. Brühl, M. Hanke, M.S. Vogelius, A direct impedance tomography algorithm for locating small inhomogeneities. Numer. Math. **93**(4), 635–654 (2003)
61. D. Bucur, G. Buttazzo, *Variational Methods in Shape Optimization Problems*. Progress in Nonlinear Differential Equations and their Applications, vol. 65 (Birkhäuser Boston, Inc., Boston, 2005)
62. M. Burger, A level set method for inverse problems. Inverse Probl. **17**, 1327–1356 (2001)
63. M. Burger, B. Hackl, W. Ring, Incorporating topological derivatives into level set methods. J. Comput. Phys. **194**(1), 344–362 (2004)
64. R.H. Burns, F.R.E. Crossley, Kinetostatic synthesis of flexible link mechanisms. ASME-Paper No. 68- Mech-36 (1964)
65. A.P. Calderón, On an inverse boundary value problem. Comput. Appl. Math. **25**(2–3), 133–138 (2006). Reprinted from the Seminar on Numerical Analysis and its Applications to Continuum Physics, Sociedade Brasileira de Matemática, Rio de Janeiro, 1980

66. D.E. Campeão, S.M. Giusti, A.A. Novotny, Topology design of plates consedering different volume control methods. Eng. Comput. **31**(5), 826–842 (2014)
67. A. Canelas, A. Laurain, A.A. Novotny, A new reconstruction method for the inverse potential problem. J. Comput. Phys. **268**, 417–431 (2014)
68. A. Canelas, A. Laurain, A.A. Novotny, A new reconstruction method for the inverse source problem from partial boundary measurements. Inverse Probl. **31**(7), 075009 (2015)
69. A. Canelas, A.A. Novotny, J.R. Roche, A new method for inverse electromagnetic casting problems based on the topological derivative. J. Comput. Phys. **230**, 3570–3588 (2011)
70. A. Canelas, A.A. Novotny, J.R. Roche, Topology design of inductors in electromagnetic casting using level-sets and second order topological derivatives. Struct. Multidiscip. Optim. **50**(6), 1151–1163 (2014)
71. Y. Capdeboscq, M.S. Vogelius, A general representation formula for boundary voltage perturbations caused by internal conductivity inhomogeneities of low volume fraction. Math. Model. Numer. Anal. **37**(1), 159–173 (2003)
72. Y. Capdeboscq, M.S. Vogelius, Optimal asymptotic estimates for the volume of internal inhomogeneities in terms of multiple boundary measurements. Math. Model. Numer. Anal. **37**(2), 227–240 (2003)
73. G. Cardone, S.A. Nazarov, J. Sokolowski, Asymptotic analysis, polarization matrices, and topological derivatives for piezoelectric materials with small voids. SIAM J. Control Optim. **48**(6), 3925–3961 (2010)
74. E.L. Cardoso, J.S.O. Fonseca, Strain energy maximization approach to the design of fully compliant mechanisms using topology optimization. Lat. Am. J. Solids Struct. **1**, 263–275 (2004)
75. A. Carpio, M.-L. Rapún, Solving inhomogeneous inverse problems by topological derivative methods. Inverse Probl. **24**(4), 045014 (2008)
76. F. Caubet, C. Conca, M. Godoy, On the detection of several obstacles in 2D Stokes flow: topological sensitivity and combination with shape derivatives. Inverse Probl. Imaging **10**(2), 327–367 (2016)
77. J. Céa, S. Garreau, Ph Guillaume, M. Masmoudi, The shape and topological optimizations connection. Comput. Methods Appl. Mech. Eng. **188**(4), 713–726 (2000)
78. D.J. Cedio-Fengya, S. Moskow, M.S. Vogelius, Identification of conductivity imperfections of small diameter by boundary measurements. Continuous dependence and computational reconstruction. Inverse Probl. **14**(3), 553–595 (1998)
79. I. Chikichev, B.B. Guzina, Generalized topological derivative for the Navier equation and inverse scattering in the time domain. Comput. Methods Appl. Mech. Eng. **194**, 4467–4484 (2008)
80. M. Dalla Riva, P. Musolino, S.V. Rogosin, Series expansions for the solution of the Dirichlet problem in a planar domain with a small hole. Asymptot. Anal. **66**(3–4), 339–361 (2015)
81. J. Rocha de Faria, A.A. Novotny, On the second order topologial asymptotic expansion. Struct. Multidiscip. Optim. **39**(6), 547–555 (2009)
82. M.C. Delfour, Topological derivative: a semidifferential via the Minkowski content. J. Convex Anal. **25**, 957–982 (2018)
83. M.C. Delfour, J.P. Zolésio, *Shapes and Geometries. Advances in Design and Control*. Society for Industrial and Applied Mathematics (SIAM) (Philadelphia, 2001)
84. N. Dominguez, V. Gibiat, Non-destructive imaging using the time domain topological energy method. Ultrasonics **50**(3), 367–372 (2010)
85. A. Drogoul, G. Aubert, The topological gradient method for semi-linear problems and application to edge detection and noise removal. Inverse Probl. Imaging **10**(1), 51–86 (2016)
86. X. Du, W. Chen, Sequential optimization and reliability assessment method for efficient probabilistic design. ASME J. Mech. Des. **126**(2), 225–233 (2004)
87. X. Duan, F. Li, Material distribution resembled level set method for optimal shape design of stokes flow. Appl. Math. Comput. **266**, 21–30 (2015)
88. P. Dunning, H. Kim, Robust topology optimization: minimization of expected and variance of compliance. Am. Inst. Aeronaut. Astronaut. J. **51**(11), 2656–2664 (2013)

89. C. Eck, J. Jarusek, J. Stará, Normal compliance contact models with finite interpenetration. Arch. Rat. Mech. Anal. **208**, 25–57 (2013)
90. I. Ekeland, R. Témam, *Convex Analysis and Variational Problems*. Volume 28 of Classics in Applied Mathematics(english edition) (Society for Industrial and Applied Mathematics (SIAM), Philadelphia, PA, 1999) (Translated from the French)
91. H. Emmendoerfer Jr., E.A. Fancello, A level set approach for topology optimization with local stress constraints. Int. J. Numer. Methods Eng. **99**, 129–156 (2014)
92. H.A. Eschenauer, V.V. Kobelev, A. Schumacher, Bubble method for topology and shape optmization of structures. Struct. Optim. **8**(1), 42–51 (1994)
93. H.A. Eschenauer, N. Olhoff, Topology optimization of continuum structures: a review. Appl. Mech. Rev. **54**(4), 331–390 (2001)
94. J.D. Eshelby, The determination of the elastic field of an ellipsoidal inclusion, and related problems. Proc. R. Soc.: Sect. A **241**, 376–396 (1957)
95. J.D. Eshelby, The elastic field outside an ellipsoidal inclusion, and related problems. Proc. R. Soc.: Sect. A **252**, 561–569 (1959)
96. E.A. Fancello, Topology optimization of minimum mass design considering local failure constraints and contact boundary conditions. Struct. Multidiscip. Optim. **32**, 229–240 (2006)
97. G.R. Feijóo, A new method in inverse scattering based on the topological derivative. Inverse Probl. **20**(6), 1819–1840 (2004)
98. R.A. Feijóo, A.A. Novotny, E. Taroco, C. Padra, The topological derivative for the Poisson's problem. Math. Models Methods Appl. Sci. **13**(12), 1825–1844 (2003)
99. L. Fernandez, A.A. Novotny, R. Prakash, Noniterative reconstruction method for an inverse potential problem modeled by a modified Helmholtz equation. Numer. Funct. Anal. Optim. **39**(9), 937–966 (2018)
100. L. Fernandez, A.A. Novotny, R. Prakash, Topological asymptotic analysis of an optimal control problem modeled by a coupled system. Asymptot. Anal. **109**, 1–26 (2018)
101. A.D. Ferreira, A.A. Novotny, A new non-iterative reconstruction method for the electrical impedance tomography problem. Inverse Probl. **33**(3), 035005 (2017)
102. G. Frémiot, W. Horn, A. Laurain, M. Rao, J. Sokołowski, On the analysis of boundary value problems in nonsmooth domains. Diss. Math. (Rozpr. Mat.) **462**, 149 (2009)
103. A. Friedman, M.S. Vogelius, Identification of small inhomogeneities of extreme conductivity by boundary measurements: a theorem on continuous dependence. Arch. Rat. Mech. Anal. **105**(4), 299–326 (1989)
104. J.F. Funes, J.M. Perales, M.L. Rapún, J.M. Manuel Vega, Defect detection from multi-frequency limited data via topological sensitivity. J. Math. Imaging Vis. **55**, 19–35 (2016)
105. S. Garreau, Ph Guillaume, M. Masmoudi, The topological asymptotic for PDE systems: the elasticity case. SIAM J. Control Optim. **39**(6), 1756–1778 (2001)
106. P. Germain, Q.S. Nguyen, P. Suquet, Continuum thermodynamics. J. Appl. Mech. Trans. ASME **50**(4), 1010–1020 (1983)
107. S.M. Giusti, Z. Mróz, J. Sokolowski, A.A. Novotny, Topology design of thermomechanical actuators. Struct. Multidiscip. Optim. **55**, 1575–1587 (2017)
108. S.M. Giusti, A.A. Novotny, E.A. de Souza Neto, Sensitivity of the macroscopic response of elastic microstructures to the insertion of inclusions. Proc. R. Soc. A: Math. Phys. Eng. Sci. **466**, 1703–1723 (2010)
109. S.M. Giusti, A.A. Novotny, E.A. de Souza Neto, R.A. Feijóo, Sensitivity of the macroscopic elasticity tensor to topological microstructural changes. J. Mech. Phys. Solids **57**(3), 555–570 (2009)
110. S.M. Giusti, A.A. Novotny, E.A. de Souza Neto, R.A. Feijóo, Sensitivity of the macroscopic thermal conductivity tensor to topological microstructural changes. Comput. Methods Appl. Mech. Eng. **198**(5–8), 727–739 (2009)
111. S.M. Giusti, A.A. Novotny, J.E. Muñoz Rivera, J.E. Esparta Rodriguez, Strain energy change to the insertion of inclusions associated to a thermo-mechanical semi-coupled system. Int. J. Solids Struct. **50**(9), 1303–1313 (2013)

112. S.M. Giusti, A.A. Novotny, J. Sokołowski, Topological derivative for steady-state orthotropic heat diffusion problem. Struct. Multidiscip. Optim. **40**(1), 53–64 (2010)
113. S.M. Giusti, J. Stebel, J. Sokołowski, On topological derivatives for contact problems in elasticity. Jounal Optim. Theory Appl. **165**, 279–294 (2015)
114. R. Glowinski, J.-L. Lions, R. Trémolières, *Numerical Analysis of Variational Inequalities*. Volume 8 of Studies in Mathematics and its Applications (North-Holland Publishing Co., Amsterdam-New York, 1981) (Translated from the French)
115. W.A. Gross, The second fundamental problem of elasticity applied to a plane circular ring. Zeitschrift fur Angewandte Mathematik und Physik **8**, 71–73 (1957)
116. P. Guillaume, K. Sid Idris, Topological sensitivity and shape optimization for the Stokes equations. SIAM J. Control Optim. **43**(1), 1–31 (2004)
117. P.H. Guillaume, M. Hassine, Removing holes in topological shape optimization. ESAIM: Control Optim. Calc. Var. **14**(1), 160–191 (2008)
118. X. Guo, W. Bai, W. Zhang, X. Gao, Confidence structural robust design and optimization under stiffness and load uncertainties. Comput. Methods Appl. Mech. Eng. **198**(41–44), 3378–3399 (2009)
119. B.B. Guzina, M. Bonnet, Small-inclusion asymptotic of misfit functionals for inverse problems in acoustics. Inverse Probl. **22**(5), 1761–1785 (2006)
120. B.B. Guzina, F. Pourahmadian, Why the high-frequency inverse scattering by topological sensitivity may work. Proc. R. Soc. A: Math. Phys. Eng. Sci. **471**(2179), 20150187 (2015)
121. B.B. Guzina, M. Bonnet, Topological derivative for the inverse scattering of elastic waves. Q. J. Mech. Appl. Math. **57**(2), 161–179 (2004)
122. A. Haraux, How to differentiate the projection on a convex set in Hilbert space. Some applications to variational inequalities. J. Math. Soc. Jpn. **29**(4), 615–631 (1977)
123. J. Haslinger, O. Vlach, Signorini problem with a solution dependent coefficient of friction (model with given friction): approximation and numerical realization. Appl. Math. **50**(2), 153–171 (2005)
124. A. Henrot, M. Pierre, *Variation et Optimisation De Formes*, vol. 48. Mathématiques et Applications (Springer, Heidelberg, 2005)
125. D. Hilding, A. Klarbring, J. Petersson, Optimization of structures in unilateral contact. Appl. Mech. Rev. **52**, 139–160 (1999)
126. M. Hintermüller, Fast level set based algorithms using shape and topological sensitivity. Control Cybern. **34**(1), 305–324 (2005)
127. M. Hintermüller, A. Laurain, Electrical impedance tomography: from topology to shape. Control Cybern. **37**(4), 913–933 (2008)
128. M. Hintermüller, A. Laurain, Multiphase image segmentation and modulation recovery based on shape and topological sensitivity. J. Math. Imaging Vis. **35**, 1–22 (2009)
129. M. Hintermüller, A. Laurain, A shape and topology optimization technique for solving a class of linear complementarity problems in function space. Comput. Optim. Appl. **46**(3), 535–569 (2010)
130. M. Hintermüller, A. Laurain, A.A. Novotny, Second-order topological expansion for electrical impedance tomography. Adv. Comput. Math. **36**(2), 235–265 (2012)
131. HistoricBridges, Hercilio Luz Bridge, http://historicbridges.org/bridges/browser/?bridgebrowser=brazil/pontehercilioluz/. Accessed 28 Sept 2016
132. I. Hlaváček, A.A. Novotny, J. Sokołowski, A. Żochowski, On topological derivatives for elastic solids with uncertain input data. J. Optim. Theory Appl. **141**(3), 569–595 (2009)
133. L.L. Howell, S.P. Magleby, B.M. Olsen, *Handbook of Compliant Mechanisms* (Wiley, Chichester, UK, 2013)
134. A.M. Il'in, *Matching of Asymptotic Expansions of Solutions of Boundary Value Problems*. Volume 102 of Translations of Mathematical Monographs (American Mathematical Society, Providence, 1992) (Translated from the Russian by V. V. Minachin)
135. V. Isakov, *Inverse Source Problems* (American Mathematical Society, Providence, Rhode Island, 1990)
136. V. Isakov, *Inverse Problems for Partial Diferential Equations* (Springer, New York, 1998)

137. V. Isakov, *Inverse Problems for Partial Differential Equations*. Applied Mathematical Sciences, vol. 127 (Springer, New York, 2006)
138. V. Isakov, S. Leung, J. Qian, A fast local level set method for inverse gravimetry. Commun. Comput. Phys. **10**(4), 1044–1070 (2011)
139. L. Jackowska-Strumiłło, J. Sokołowski, A. Żochowski, The topological derivative method and artificial neural networks for numerical solution of shape inverse problems. Technical Report 3739, INRIA-Lorraine (1999)
140. L. Jackowska-Strumiłło, J. Sokołowski, A. Żochowski, A. Henrot, On numerical solution of shape inverse problems. Comput. Optim. Appl. **23**(2), 231–255 (2002)
141. M. Jleli, B. Samet, G. Vial, Topological sensitivity analysis for the modified Helmholtz equation under an impedance condition on the boundary of a hole. J. Math. Pures Appl. **103**, 557–574 (2015)
142. M. Kachanov, B. Shafiro, I. Tsukrov, *Handbook of Elasticity Solutions* (Kluwer Academic Publishers, Dordrecht, 2003)
143. I.V. Kamotski, S.A. Nazarov, Spectral problems in singular perturbed domains and self adjoint extensions of differential operators. Trudy St.-Petersburg Mat. Obshch. **6**, 151–212 (1998) (English transl. in Proceedings of the St. Petersburg Mathematical Society, 6:127–181, 2000. Amer. Math. Soc. Transl. (Ser. 2), 199, Amer. Math. Soc., Providence, RI)
144. A.M. Khludnev, V.A. Kovtunenko, *Analysis of Cracks in Solids* (WIT Press, Southampton-Boston, 2000)
145. A.M. Khludnev, A.A. Novotny, J. Sokołowski, A. Żochowski, Shape and topology sensitivity analysis for cracks in elastic bodies on boundaries of rigid inclusions. J. Mech. Phys. Solids **57**(10), 1718–1732 (2009)
146. A.M. Khludnev, J. Sokołowski, *Modelling and Control in Solid Mechanics* (Birkhauser, Basel-Boston-Berlin, 1997)
147. A.M. Khludnev, J. Sokołowski, Griffith formulae for elasticity systems with unilateral conditions in domains with cracks. Eur. J. Mech. A/Solids **19**, 105–119 (2000)
148. A.M. Khludnev, J. Sokołowski, On differentation of energy functionals in the crack theory with possible contact between crack faces. J. Appl. Math. Mech. **64**(3), 464–475 (2000)
149. N. Kikuchi, S. Nishiwaki, J.S.O. Fonseca, E.C.N. Silva, Design optimization method for compliant mechanisms and material microstructure. Comput. Methods Appl. Mech. Eng. **151**(3–4), 401–417 (1998)
150. V. Kobelev, Bubble-and-grain method and criteria for optimal positioning inhomogeneities in topological optimization. Struct. Multidiscip. Optim. **40**(1–6), 117–135 (2010)
151. P.I. Kogut, G. Leugering, *Optimal Control Problems for Partial Differential Equations on Reticulated Domains: Approximation and Asymptotic Analysis* (Springer, Berlin, 2011)
152. R.V. Kohn, H. Shen, M.S. Vogelius, M.I. Weinstein, Cloaking via change of variables in electric impedance tomography. Inverse Probl. **24**(1), 015016 (2008)
153. R.V. Kohn, M.S. Vogelius, Determining conductivity by boundary measurements. Commun. Pure Appl. Math. **37**(3), 289–298 (1984)
154. R.V. Kohn, M.S. Vogelius, Identification of an unknow conductivity by means of measurements at the boundary. Inverse Probl. **14**, 113–123 (1984)
155. R.V. Kohn, M.S. Vogelius, Relaxation of a variational method for impedance computed tomography. Commun. Pure Appl. Math. **40**(6), 745–777 (1987)
156. P. Kurasov, A. Posilicano, Finite speed of propagation and local boundary conditions for wave equations with point interactions. Proc. Am. Math. Soc. **133**(10), 3071–3078 (2005)
157. I. Larrabide, R.A. Feijóo, A.A. Novotny, E. Taroco, Topological derivative: a tool for image processing. Comput. Struct. **86**(13–14), 1386–1403 (2008)
158. I. Lasiecka, *Mathematical Control Theory of Coupled PDEs*. Volume 75 of CBMS-NSF Regional Conference Series in Applied Mathematics (Society for Industrial and Applied Mathematics (SIAM), Philadelphia, 2002)
159. A. Laurain, M. Hintermüller, M. Freiberger, H. Scharfetter, Topological sensitivity analysis in fluorescence optical tomography. Inverse Probl. **29**(2), 025003,30 (2013)

160. A. Laurain, S.A. Nazarov, J. Sokolowski, Singular perturbations of curved boundaries in three dimensions. The spectrum of the Neumann Laplacian. Zeitschrift für Analysis und ihre Anwendungen. J. Anal. Appl. **30**(2), 145–180 (2011)
161. C. Le, J. Norato, T. Bruns, Stress-based topology optimization for continua. Struct. Multidiscip. Optim. **41**, 605–620 (2010)
162. E. Lee, H.C. Gea, A strain based topology optimization method for compliant mechanism design. Struct. Multidiscip. Optim. **49**, 199–207 (2014)
163. A. Leitão, J. Baumeister, *Topics in Inverse Problems* (IMPA Mathematical Publications, Rio de Janeiro, 2005)
164. G. Leugering, S.A. Nazarov, F. Schury, M. Stingl, The eshelby theorem and application to the optimization of an elastic patch. SIAM J. Appl. Math. **72**(2), 512–534 (2012)
165. G. Leugering, A.A. Novotny, G. Perla-Menzala, J. Sokołowski, Shape sensitivity analysis of a quasi-eletrostatic piezoelectric system in multilayered media. Math. Methods Appl. Sci. **33**(17), 2118–2131 (2010)
166. G. Leugering, A.A. Novotny, G. Perla-Menzala, J. Sokołowski, On shape optimization for an evolution coupled system. Appl. Math. Optim. **64**, 441–466 (2011)
167. G. Leugering, J. Sokołowski, Topological derivatives for elliptic problems on graphs. Control Cybern. **37**, 971–998 (2008)
168. G. Leugering, J. Sokołowski, A. Żochowski, Control of crack propagation by shape-topological optimization. Discret. Contin. Dyn. Syst. Ser. A **35**(6), 2625–2657 (2015)
169. T. Lewiński, J. Sokołowski, Optimal shells formed on a sphere. the topological derivative method. Technical Report 3495, INRIA-Lorraine (1998)
170. T. Lewiński, J. Sokołowski, Topological derivative for nucleation of non-circular voids. The Neumann problem, in *Differential Geometric Methods in the Control of Partial Differential Equations (Boulder, CO, 1999)*. Volume 268 of Contemporary Mathematics (American Mathematical Society, Providence, 2000), pp. 341–361
171. T. Lewiński, J. Sokołowski, Energy change due to the appearance of cavities in elastic solids. Int. J. Solids Struct. **40**(7), 1765–1803 (2003)
172. T. Lewiński, J. Sokołowski, A. Żochowski, Justification of the bubble method for the compliance minimization problems of plates and spherical shells, in *In 3rd World Congress of Structural and Multidisciplinary Optimization (WCSMO-3)* (Buffalo/Niagara Falls, New York, 1999)
173. T. Lewiński, J.J. Telega, *Plates, Laminates and Shells*. Volume 52 of Series on Advances in Mathematics for Applied Sciences (World Scientific Publishing Co., Inc, River, Edge, 2000) (Asymptotic analysis and homogenization)
174. Y. Li, K. Saitou, N. Kikuchi, Topology optimization of thermally actuated compliant mechanisms considering time-transient effect. Finite Elem. Anal. Des. **40**, 1317–1331 (2004)
175. A. Logg, K.-A. Mardal, G.N. Wells, *Automated Solution of Differential Equations by the Finite Element Method. The FEniCS book*. Volume 84, Lecture Notes in Computational Science and Engineering (Springer, Berlin, 2012)
176. C.G. Lopes, R.B. dos Santos, A.A. Novotny, Topological derivative-based topology optimization of structures subject to multiple load-cases. Lat. Am. J. Solids Struct. **12**, 834–860 (2015)
177. C.G. Lopes, A.A. Novotny, Topology design of compliant mechanisms with stress constraints based on the topological derivative concept. Struct. Multidiscip. Optim. **54**(4), 737–746 (2016)
178. C.G. Lopes, R.B. Santos, A.A. Novotny, J. Sokołowski, Asymptotic analysis of variational inequalities with applications to optimum design in elasticity. Asymptot. Anal. **102**, 227–242 (2017)
179. J. Luo, Z. Luo, S. Chen, L. Tong, M. Yu Wang, A new level set method for systematic design of hinge-free compliant mechanisms. Comput. Methods Appl. Mech. Eng. **198**, 318–331 (2008)
180. A.I. Lurie, *Theory of Elasticity* (Springer, Berlin, 2005)
181. T.J. Machado, J.S. Angelo, A.A. Novotny, A new one-shot pointwise source reconstruction method. Math. Methods Appl. Sci. **40**(15), 1367–1381 (2017)

182. N.D. Mankame, G.K. Ananthasuresh, Topology optimization for synthesis of contact-aided compliant mechanisms using regularized contact modeling. Comput. Struct. **82**, 1267–1290 (2004)
183. M. Masmoudi, J. Pommier, B. Samet, The topological asymptotic expansion for the Maxwell equations and some applications. Inverse Probl. **21**(2), 547–564 (2005)
184. W.G. Mazja, S.A. Nasarow, B.A. Plamenewski, *Asymptotics of Solutions to Elliptic Boundary-value Problems Under a Singular Perturbation of the Domain* (Tbilisi University, Tbilisi, 1981). (in Russian)
185. V. Maz'ya, S.A. Nazarov, B.A. Plamenevskij, *Asymptotic Theory of Elliptic Boundary Value Problems in Singularly Perturbed Domains*, vol. I. Volume 111 of Operator Theory: Advances and Applications (Birkhäuser Verlag, Basel, 2000) (Translated from the German by Georg Heinig and Christian Posthoff)
186. V. Maz'ya, S.A. Nazarov, B.A. Plamenevskij, *Asymptotic Theory of Elliptic Boundary Value Problems in Singularly Perturbed Domains*, vol. II. Volume 112 of Operator Theory: Advances and Applications (Birkhäuser Verlag, Basel, 2000) (Translated from the German by Plamenevskij)
187. V.G. Maz'ya, S.A. Nazarov, B.A. Plamenevskij, *Asymptotische Theorie Elliptischer Randwertaufgaben in Singulär Gestörten Gebieten*, vol. 1. (Akademie-Verlag, Berlin, 1991) (English transl.: Asymptotic theory of elliptic boundary value problems in singularly perturbed domains, vol. 1, Basel: Birkhäuser Verlag, 2000)
188. V.G. Maz'ya, S.A. Nazarov, B.A. Plamenevskij, *Asymptotic Theory of Elliptic Boundary Value Problems in Singularly Perturbed Domains*, vol. I. Volume 111 of Operator Theory: Advances and Applications (Birkhäuser Verlag, Basel, 2000) (Translated from the German by Georg Heinig and Christian Posthoff)
189. V.N. Melnik, Existence and uniqueness theorems of the generalized solution for a class of non-stationary problem of coupled electroelasticity. Sov. Math. Izv. VUZ. Math. **35**(4), 24–32 (1991)
190. C.G. Méndez, J.M. Podestá, O. Lloberas-Valls, S. Toro, A.E. Huespe, J. Oliver, Computational material design for acoustic cloaking. Int. J. Numer. Methods Eng. **112**(10), 1353–1380 (2017)
191. L.R. Meneghelli, E.L. Cardoso, Design of compliant mechanisms with stress constraints using topology optimization. Optim. Struct. Compon. Adv. Struct. Mater. **43**, 35–48 (2013)
192. D. Mercier, S. Nicaise, Existence, uniqueness and regularity results for piezoelectric systems. SIAM J. Math. Anal. **37**, 651–672 (2005)
193. J.C. Michel, H. Moulinec, P. Suquet, Effective properties of composite materials with periodic microstructure: a computational approach. Comput. Methods Appl. Mech. Eng. **172**(1–4), 109–143 (1999)
194. C. Miehe, J. Schotte, J. Schröder, Computational micro-macro transitions and overall moduli in the analysis of polycrystals at large strains. Comput. Mater. Sci. **16**(1–4), 372–382 (1999)
195. F. Mignot, Contrôle dans les inéquations variationelles elliptiques. J. Funct. Anal. **22**(2), 130–185 (1976)
196. N.I. Muskhelishvili, *Some Basic Problems on the Mathematical Theory of Elasticity* (Noordhoff, Groningen, 1952)
197. M. Muszkieta, A variational approach to edge detection. Inverse Probl. Imaging **10**(2), 499–517 (2016)
198. S.A. Nazarov, *Asymptotic Expansions of Eigenvalues* (Leningrad University, Leningrad, 1987)
199. S.A. Nazarov, Self-adjoint elliptic boundary-value problems. the polynomial property and formally positive operators. Probl. Mat. Anal. **16**, 167–192 (1997) (English transl.: J. Math. Sci. 92:4338–4353, 1998)
200. S.A. Nazarov, Asymptotic conditions at a point, selfadjoint extensions of operators, and the method of matched asymptotic expansions. Am. Math. Soc. Transl. **198**, 77–125 (1999)
201. S.A. Nazarov, *Asymptotic Theory of Thin Plates and Rods. Volume 1: Dimension Reduction and Integral Estimates* (Nauchnaya Kniga, Novosibirsk, 2001)
202. S.A. Nazarov, Elasticity polarization tensor, surface enthalpy and Eshelby theorem. Probl. Mat. Analiz. **41**, 3–35 (2009) (English transl.: Journal of Math. Sci. 159(1–2):133–167, 2009)

203. S.A. Nazarov, The Eshelby theorem and a problem on an optimal patch. Algebra i Analiz. **21**(5), 155–195 (2009) (English transl.: St. Petersburg Math. 21(5):791–818, 2009)
204. S.A. Nazarov, B.A. Plamenevskij, *Elliptic Problems in Domains with Piecewise Smooth Boundaries*, vol. 13. de Gruyter Expositions in Mathematics (Walter de Gruyter & Co., Berlin, 1994)
205. S.A. Nazarov, J. Sokołowski, Asymptotic analysis of shape functionals. J. Math. Pures Appl. **82**(2), 125–196 (2003)
206. S.A. Nazarov, J. Sokołowski, Selfadjoint extensions for the elasticity system in shape optimization. Bull. Pol. Acad. Sci. Math. **52**(3), 237–248 (2004)
207. S.A. Nazarov, J. Sokolowski, Modeling of topology variations in elasticity, in *System Modeling and Optimization*. Volume 166 of IFIP International Federation for Information Processing (Kluwer Academic Publishers, Boston, 2005), pp. 147–158
208. S.A. Nazarov, J. Sokołowski, Self-adjoint extensions of differential operators and exterior topological derivatives in shape optimization. Control Cybern. **34**(3), 903–925 (2005)
209. S.A. Nazarov, J. Sokołowski, Self-adjoint extensions for the Neumann Laplacian and applications. Acta Math. Sin. (Engl. Ser.) **22**(3), 879–906 (2006)
210. S.A. Nazarov, J. Sokolowski, Shape sensitivity analysis of eigenvalues revisited. Control Cybern. **37**(4), 999–1012 (2008)
211. S.A. Nazarov, J. Sokołowski, Spectral problems in the shape optimisation. Singul. Bound. Perturbations. Asymptot. Anal. **56**(3–4), 159–204 (2008)
212. S.A. Nazarov, J. Sokołowski, On asymptotic analysis of spectral problems in elasticity. Lat. Am. J. Solids Struct. **8**, 27–54 (2011)
213. S.A. Nazarov, J. Sokołowski, M. Specovius-Neugebauer, Polarization matrices in anisotropic heterogeneous elasticity. Asymptot. Anal. **68**(4), 189–221 (2010)
214. A.A. Novotny, Sensitivity of a general class of shape functional to topological changes. Mech. Res. Commun. **51**, 1–7 (2013)
215. A.A. Novotny, R.A. Feijóo, C. Padra, E. Taroco, Topological sensitivity analysis. Comput. Methods Appl. Mech. Eng. **192**(7–8), 803–829 (2003)
216. A.A. Novotny, R.A. Feijóo, C. Padra, E. Taroco, Topological derivative for linear elastic plate bending problems. Control Cybern. **34**(1), 339–361 (2005)
217. A.A. Novotny, R.A. Feijóo, E. Taroco, C. Padra, Topological sensitivity analysis for three-dimensional linear elasticity problem. Comput. Methods Appl. Mech. Eng. **196**(41–44), 4354–4364 (2007)
218. A.A. Novotny, V. Sales, Energy change to insertion of inclusions associated with a diffusive/convective steady-state heat conduction problem. Math. Methods Appl. Sci. **39**(5), 1233–1240 (2016)
219. A.A. Novotny, J. Sokołowski, *Topological Derivatives in Shape Optimization*. Interaction of Mechanics and Mathematics (Springer, Berlin, 2013)
220. A.A. Novotny, J. Sokołowski, E.A. de Souza Neto, Topological sensitivity analysis of a multi-scale constitutive model considering a cracked microstructure. Math. Methods Appl. Sci. **33**(5), 676–686 (2010)
221. M. Otomori, T. Yamada, K. Izui, S. Nishiwaki, Matlab code for a level set-based topology optimization method using a reaction diffusion equation. Struct. Multidiscip. Optim. **51**(5), 1159–1172 (2015)
222. B.S. Pavlov, The theory of extensions, and explicitly solvable models. Akademiya Nauk SSSR i Moskovskoe Matematicheskoe Obshchestvo. Uspekhi Matematicheskikh Nauk **42(6(258)): 99–131**, 247 (1987)
223. J.T. Pereira, E.A. Fancello, C.S. Barcellos, Topology optimization of continuum structures with material failure constraints. Struct. Multidiscip. Optim. **26**(1–2), 50–66 (2004)
224. P. Plotnikov, J. Sokołowski, *Compressible Navier-Stokes Equations*. Theory and Shape Optimization (Springer, Basel, 2012)
225. S.S. Rocha, A.A. Novotny, Obstacles reconstruction from partial boundary measurements based on the topological derivative concept. Struct. Multidiscip. Optim. **55**(6), 2131–2141 (2017)

226. W.M. Rubio, E.C.N. Silva, S. Nishiwaki, Design of compliant mechanisms considering thermal effect compensation and topology optimization. Finite Elem. Anal. Des. **46**, 1049–1060 (2010)
227. L.F.N. Sá, R.C.R. Amigo, A.A. Novotny, E.C.N. Silva, Topological derivatives applied to fluid flow channel design optimization problems. Struct. Multidiscip. Optim. **54**(2), 249–264 (2016)
228. V. Sales, A.A. Novotny, J.E. Muñoz Rivera, Energy change to insertion of inclusions associated with the reissner-mindlin plate bending model. Int. J. Solids Struct. **59**, 132–139 (2013)
229. B. Samet, Topological sensitivity analysis with respect to a small hole located at the boundary of the domain. Asymptot. Anal. **66**(1), 35–49 (2010)
230. B. Samet, S. Amstutz, M. Masmoudi, The topological asymptotic for the Helmholtz equation. SIAM J. Control Optim. **42**(5), 1523–1544 (2003)
231. E. Sanchez-Palencia, *Non-homogeneous Media and Vibration Theory*, vol. 127. Lecture Notes in Physics (Springer, Berlin, 1980)
232. M. Schneider, H. Andrä, The topological gradient in anisotropic elasticity with an eye towards lightweight design. Math. Methods Appl. Sci. **37**, 1624–1641 (2014)
233. A. Schumacher, Topologieoptimierung von bauteilstrukturen unter verwendung von lochpositionierungkriterien. Ph.D. Thesis, Universität-Gesamthochschule-Siegen, Siegen - Germany (1995)
234. O. Sigmund, On the design of compliant mechanisms using topology optimization. Mech. Struct. Mach.: Int. J. **25**(4), 493–524 (1997)
235. E.C.N. Silva, N. Kikuchi, Design of piezoelectric tranducers using topology optimization. Smart Mater. Struct. **8**, 350–364 (1999)
236. J. Sokoł owski, Displacement derivatives in shape optimization of thin shells, in *Optimization Methods in Partial Differential Equations (South Hadley, MA, 1996)*. Volume 209 of Contemporary Mathematics (American Mathematical Society, Providence, 1997), pp. 247–266
237. J. Sokołowski, Sensitivity analysis of contact problems with prescribed friction. Appl. Math. Optim. **18**(2), 99–117 (1988)
238. J. Sokołowski, A. Żochowski, On the topological derivative in shape optimization. SIAM J. Control Optim. **37**(4), 1251–1272 (1999)
239. J. Sokołowski, A. Zochowski, Topological derivative for optimal control problems. Control Cybern. **28**(3), 611–626 (1999)
240. J. Sokołowski, A. Żochowski, Topological derivatives of shape functionals for elasticity systems. Mech. Struct. Mach. **29**(3), 333–351 (2001)
241. J. Sokołowski, A. Żochowski, Optimality conditions for simultaneous topology and shape optimization. SIAM J. Control Optim. **42**(4), 1198–1221 (2003)
242. J. Sokołowski, A. Żochowski, Modelling of topological derivatives for contact problems. Numer. Math. **102**(1), 145–179 (2005)
243. J. Sokołowski, A. Żochowski, Topological derivatives in plane elasticity, in *23rd IFIP TC 7* Conference on System Modeling and Optimization (Berlin, Germany, 2009)
244. J. Sokołowski, J.-P. Zolésio, Dérivée par rapport au domaine de la solution d'un problème unilatéral [shape derivative for the solutions of variational inequalities]. C. R. Acad. Sci. Paris Sér. I Math. **301**(4), 103–106 (1985)
245. J. Sokołowski, J.P. Zolésio, *Introduction to Shape Optimization - Shape Sensitivity Analysis* (Springer, Berlin, 1992)
246. N. Strömberg, A. Klarbring, Topology optimization of structures in unilateral contact. Struct. Multidiscip. Optim. **41**(1), 57–64 (2010)
247. R. Tokmashev, A. Tixier, B. Guzina, Experimental validation of the topological sensitivity approach to elastic-wave imaging. Inverse Probl. **29**, 125005 (2013)
248. A.J. Torii, A.A. Novotny, R.B. Santos, Robust compliance topology optimization based on the topological derivative concept. Int. J. Numer. Methods Eng. **106**(11), 889–903 (2016)
249. P. Tricarico, Global gravity inversion of bodies with arbitrary shape. Geophys. J. Int. **195**(1), 260–275 (2013)

250. I. Tsukrov, M. Kachanov, Effective moduli of an anisotropic material with elliptical holes of arbitrary orientational distribution. Int. J. Solids Struct. **37**(41), 5919–5941 (2000)
251. J. Tu, K.K. Choi, Y.H. Park, A new study on reliability-based design optimization. ASME J. Mech. Des. **121**(4), 557–564 (1999)
252. I. Turevsky, S.H. Gopalakrishnan, K. Suresh, An efficient numerical method for computing the topological sensitivity of arbitrary-shaped features in plate bending. Int. J. Numer. Methods Eng. **79**(13), 1683–1702 (2009)
253. N. Van Goethem, A.A. Novotny, Crack nucleation sensitivity analysis. Math. Methods Appl. Sci. **33**(16), 1978–1994 (2010)
254. M.S. Vogelius, D. Volkov, Asymptotic formulas for perturbations in the electromagnetic fields due to the presence of inhomogeneities of small diameter. ESAIM Math. Model. Numer. Anal. **37**, 723–748 (2000)
255. P. Wriggers, *Computational Contact Mechanics* (Springer, Berlin, 2006)
256. M. Xavier, E.A. Fancello, J.M.C. Farias, N. Van Goethem, A.A. Novotny, Topological derivative-based fracture modelling in brittle materials: a phenomenological approach. Eng. Fract. Mech. **179**, 13–27 (2017)
257. M. Xavier, A.A. Novotny, N. Van Goethem, A simplified model of fracking based on the topological derivative concept. Int. J. Solids Struct. **139–140**, 211–223 (2018)
258. J. Zhao, C. Wang, Robust topology optimization under loading uncertainty based on linear elastic theory and orthogonal diagonalization of symmetric matrices. Comput. Methods Appl. Mech. Eng. **273**, 204–218 (2014)
259. O.C. Zienkiewicz, R.L. Taylor, *The Finite Element Method*, 5th edn. (Butterworth-Heinemann, Oxford, 2000)

The manufacturer's authorised representative in the EU is Springer Nature Customer Service Centre GmbH, Europaplatz 3, 69115 Heidelberg, Germany. If you have any concerns regarding our products, please contact ProductSafety@springernature.com

Printed and bound by CPI Group (UK) Ltd, Croydon, CR0 4YY
22/09/2026
02221062-0006